STATELY FOSSILS

A COMPREHENSIVE LOOK AT THE STATE FOSSILS AND OTHER OFFICIAL FOSSILS

STEPHEN BRUSATTE

Fossil News
Boulder, Colorado

Published by and available from
Fossil News
1185 Claremont Drive
Boulder, Colorado 80305

This book was produced entirely on Macintosh computers—a G4 Cube, an iBook, and a G3 Desktop. The text was set in 12 pt. AGaramond and the headings are in 18 pt. Capitals. Adobe Pagemaker software was used for the layout.

Book and cover design by Lynne M. Clos

Printed in the U.S.A.
November, 2002

ISBN 0-9724416-2-X

Library of Congress Control Number: 2002115964

Dedicated to Jack Sepkoski and Nick Hotton—your wisdom extends farther than you ever knew.

CONTENTS

FOREWORD

So there I was, sitting on the curb in front of our Park Forest, Illinois home one late summer afternoon, back in the year Maris hit 61. That day I found my first fossil, a cylindrically-shaped bit of whitish stone encircled with grooves, reminding me of a broken screw. My avocational interests in paleontology stemmed from what happened next—a simple thing, really, just asking my Dad what it was. He not only identified it, but also explained its significance. This became a major milestone in my life.

In explaining it was something called a "crinoid," my Dad opened a remarkable window into the deep past, a portal into "prehistory" that began long before the time of Stonehenge or the age of Ancient Egypt and its dusty mummies and pyramids, spanning periods of time that seemed incomprehensibly long ago. I realized that fossils such as my crinoid stem permitted curious-minded souls to catch fleeting glimpses of a world vastly different from today's, a world before the time of modern humans and our inventions. I listened to words that day which I could barely comprehend, such as "Paleozoic," "Silurian," and "echinoderm," yet I thirsted for more. As I still do.

Fossil collecting became a 1960s passion of mine. Though I was still in grade school and had never collected fossils *in situ* or in the field, I had become a self-made "expert" on fossils after absorbing my favorite book of the time devoted to fossils—*Fossils: A Guide to Prehistoric Life* (A Golden Nature Guide), New York: Golden Press, 1962, by Frank H.T. Rhodes, Herbert S. Zim, and Paul R. Shaffer, that advertised on the front cover its "481 illustrations in color." That book is still nestled on my bookshelf, its pages well worn from the many hours of youthful fossil-hunting excursions it was carried on.

We were lucky to have fossiliferous rocks in the immediate vicinity of places we lived, especially during the years between 1964 and 1968. Initially, my younger brother (who is now a petroleum geologist) and I would sit in the empty gravel parking lot of our grade school, turning over limestone rocks, searching for fossils nearly every summer day. What hauls we made! After a couple hours of picking over stones, we would cross the grassy field to our house, sunburned and sweaty, our pockets bursting with valuable "finds." Developing an uncanny sixth sense for when our best discoveries were about to be made, we found gastropods, brachiopods, cephalopod fragments, coral fragments, crinoid stems and calyxes, and even trilobite fragments. Many of our treasures were in beautiful condition. These hunks of rock from a prehistoric age fascinated us. It amazed us that nobody else had thought to comb these fossiliferous fields. I later discovered that the gravel had come from the nearby Thornton Quarry, a Niagara limestone reef that we collected in a few years later.

We spent a year and two subsequent summers in England, and found more gravel deposits laden with Mesozoic treasures right outside our apartments, there just ripe for the picking! Now we ambitiously added ammonites, belemnites, clams, sponges, sea urchins and worm tubes to our collection. Several years later, by 1970 through 1972, my parents had been cajoled into taking us to quarries, exposures, and road cuts to collect fossils both in Illinois (participating in field trips sponsored by the Illinois State Geological Survey), and also in England (Cambridge area). But we learned of the hazards of field collecting, too. While we happily hammered away on limestones, my Mom was regularly and savagely attacked by bees that were attracted to her hairspray.

As our collections grew, so did our library of paleontology books. Our knowledge of this fascinating subject flowered. We made every effort to identify what we found using the references we accumulated.

There were also late October college geology and paleontology field trips to sites in Iowa, Arkansas, and Wisconsin. Throughout, I labeled my fossils dutifully and filled two notebooks with relevant notes, even illustrating the best fossils for posterity. I still have those notebooks today. I also learned the "pleasures" of camping from these experiences, after camping in freezing conditions, one time in 18°F weather, and stomping about with frozen feet that I couldn't feel for several days afterward (no doubt partly due to the "calming" beverages we heartily consumed the night before to "thin" the blood in our veins).

My fossil collection still resides in our garage, spared from numerous spring-cleaning "downsizings." Into adulthood interests naturally shift, although my enthusiasm for paleontology, dinosaurs, and, yes, fossils has remained a constant.

It was sometime in sixth grade that I must have declared that I wanted to become a paleontologist when I grew up. Immediately concern was raised over how I would starve, as this was not a lucrative field. Paleontology was viewed as a "dead" science, unexciting because there was so little happening in the profession. There would be few jobs for college graduates. Oh, the inhumanity of it all! If only I had known how substantially the circumstances would all change in such a short time. For in short succession we experienced: the plate tectonics revolution with its reliance on fossils; the enlightening punctuated equilibrium theory of organic evolution; the dinosaur renaissance; a lively debate over the nature and causes of mass extinctions; and a host of fossil discoveries that would make our heads spin with joy and astonishment. Paleontology was enlivened, and thanks to my Dad's encouraging description of what a strange screw-shaped stony fragment was, long ago, I found it impossible to let go.

And so I embarked on a professional career in the field of chemistry, while avidly maintaining paleontological interests. After the mid-1980s, when it became virtually impossible to find the time to collect fossils, I turned to writing about paleontological subjects, and sculpting "scenes from deep time." One of the recent manifestations of this spirited hobby has been my contributions to *Fossil News: Journal of Avocational Paleontology.* This is how I came to meet a remarkable youth—Steve Brusatte!

I am uncertain how we first became acquainted, but it must have been through Internet correspondence about the time he was conducting interviews for a book he was researching. Steve and I have become good friends over the years, and we have attended lectures on "Sue" at the Field Museum of Natural History and also at the Burpee Museum of Natural History's annual "Out of the Rock Paleo-Festival". Steve and I have even co-authored a couple of articles, published in *Fossil News.* Steve is a dedicated paleontologist, and I know how much he means it when he says that he wants to become a professional paleontologist. Paleontology would benefit from having him join the ranks of the "bug hunters."

In the fall of 1998, Lynne Clos, who had recently taken the helm as editor of *Fossil News*, asked me if I could write any articles for the journal concerning any of my past fossil hunting escapades. Given that I've never collected in the same professional sense as some of her other published fossil-collecting contributors, I hesitated, replying facetiously that if she ever printed an article of mine on fossil collecting it could only mean she had become desperate for manuscripts! Then, several months later, I had a counter-suggestion for her. I recommended that a series of articles on the U.S. "state fossils" would make a great recurring piece in her magazine, although declining to pen the articles myself. We both thought that Steve would be a good candidate to write the articles. Voila! Not only were the articles written, superbly, but they have been compiled—and expanded—herein for you to enjoy.

Was there an ulterior motive? Well, this is a tale I've often wanted to read about, and I found a person to write it for me to read, without having to do all the research myself. Sounds suspicious, huh? Anyway you slice it, it's a good deal for all of us.

But enough of this. Now it's time to read about screw-shaped rocks and other "stately" fossils. Nice job, Steve, and to the rest of you—enjoy! Read on.

Allen A. Debus
Hanover Park, IL
July 7, 2001

ACKNOWLEDGMENTS

There are many people to thank for their involvement in this project, as a considerable number of individuals extended their guidance and help, and were instrumental in making this book what it is. First and foremost, I would like to thank Lynne Clos and Allen Debus. Lynne has done much of the legwork when it comes to this book, both providing me with the opportunity to write for her magazine, *Fossil News*, and in the technical production of the book. Allen must be recognized for suggesting the idea to me, critiquing my writing, and penning the excellent foreword.

Of course, I would also like to express my gratitude towards my family—my father Jim, mother Roxanne, and brothers Mike and Chris. My father has been most helpful in reviewing and correcting my writing, and supporting my interest in paleontology. My mother has been equally supportive, and both of my brothers have aided me in my quest to write this book. The support of my family cannot be expressed in words.

In the paleontological community, I also thank, in no particular order, Ken Kinman, Ray Stanford, Tim Williams, Tom Lipka, Patrick Norton, Charlie Sturm, Mike Fredericks, Klaus Westphal, Zhexi Luo, Jason Anderson, Fred Bervoets, Joanne Kluessendorf, Donald Mikulic, Paul Sereno, Gary Gaulin, Lonny Stark, Dean Walker, Fabio Marco Dalla Vecchia, Walter Alvarez, Peter Dodson, John Lanzendorf, William Hammer, Don Lessem, Adrienne Mayor, Dennis Kolata, LeGrand Smith, Tony Campagna, and Jaime Headden. The websites of Paul Heinrich and Ed Summer have also been very helpful. Also, the advice of my teacher, Joe Jakupcak, has helped mold me into a better scientist.

Finally, I would like to recognize the efforts of the late Jack Sepkoski and Nick Hotton, two of the greatest fossil researchers of all time. Both were instrumental in guiding me when I first began my paleontological quest, and I was saddened to see both of them go. This book is for you!

PREFACE

It all started with a plea from my brother. I was fourteen years old at the time, and fossils were the last thing on my mind. Instead, during the summer of 1998, I was focused on Major League Baseball's home run race and my looming entrance to high school. However, fossils literally exploded onto the scene, and have become such a part of me that we will likely remain intertwined forever.

During that summer, my youngest brother Chris, only ten, had begun working on a short book for a school competition. Both Chris and I were interested in dinosaurs and fossils as young children, but while paleontology left me as I entered grade school, it stuck with Chris. Over the years he amassed dinosaur toys, faithfully read a slew of dinosaur books, and even turned his bedroom into a miniature dinosaur museum. That "museum," which he called Dino Land, housed his impressive collection of dinosaur memorabilia.

The book Chris wanted to write concerned the contributions of famous paleontologists, and in preparation he pored over his collection of books, marking the names of the more famous scientists. Naturally, in a science as fruitful as paleontology, there is no short list of names. He soon became tired of looking, and asked his older brother for help. It may not be the most orthodox of stories, but soon I was hooked. To this day, over three years later, I am not only still fascinated with paleontology, but also devote much of my life to it.

Soon after rediscovering fossils, I began to purchase my own books, subscribe to amateur magazines, create a website, visit museums, and hunt for fossils. After a short while I proclaimed that I wanted to become the first teenager to write a book on paleontology. While I was most definitely underqualified to author a fossil volume, I persisted.

I first announced that I would pen a title on the evolution of life through the ages, presented in a simplified, easy-to-read volume that could be enjoyed by children and adults both. However, over time, I saw how difficult this was, and how much more I needed to know before I could accomplish such a task. I switched gears, deciding to write a book of essays, á la Stephen Jay Gould, on a teenager's perspective on the field. But this, too, became too burdensome, and

I soon gave it up. My book writing was not finished, though—as I would soon discover.

In the spring of 1999 I began writing for a splendid monthly magazine called *Fossil News*. Edited by a free lance paleontologist named Lynne Clos, this was a professionally-published journal created for amateur paleontologists. I was definitely an amateur, and decided to try my hand at magazine writing. I have written several articles for *Fossil News*, and still enjoy contributing.

In January of 2000, Lynne told me about a feature that she wished somebody would consider writing. When she first took over the magazine from its founder in Dash Point, Washington, *Fossil News* contained a monthly Children's Corner article. These articles, which were written in simple words and took up a single page, were meant for children, but could also be enjoyed by adults. Over time the feature became extinct, and Lynne was interested in having it return. I volunteered, renaming it the Kids' Corner. For the first few months I wrote about a variety of topics, ranging from *Archaeopteryx* to Mazon Creek. However, soon after I started, I was approached with a new idea.

Allen Debus, a self-taught dinosaur maven, is one of the more accomplished amateurs in the field. The son of a well-known professor of science, Allen is an environmental chemist by trade, but paleontology fills his heart. After I was introduced to *Dinosaur World*, an amateur magazine Allen co-edited, he and I became good friends. We corresponded over the Internet, co-authored a series of articles together, and attended many of the same lectures. Allen also recommended a new Kids' Corner topic that has changed my life.

Many states in the U. S. recognize state fossils and state dinosaurs, along with more traditional symbols such as state seal and state flag. In addition, a select number of states also have designated fossils as state stones and state gems. Allen had long been intrigued by these state fossils, both with their physical characteristics and with the human process involved in their designation. However, he didn't have the time to do any research or write articles on the topic. Instead, he recommended that I use my Kids' Corner column as a forum to discuss these state fossils. I gleefully accepted, and in May, 2000 my first piece, a one-page long article on *Allosaurus* of Utah, was published in *Fossil News*.

I continued to write about these state fossils, when Allen casually suggested that a book on state fossils would be a good idea. Although my two previous book ideas had failed, there was still a longing in my heart to complete a book before I entered college. I saw this as a good opportunity, and told Allen that I would be much interested in authoring a book on these stately fossils. I later discussed the idea with Lynne Clos, and after considering various options, we decided to produce the book ourselves. I began the volume in March, 2001, and completed the first draft in July. The finished product is before you.

Trying to complete a book while balancing school, work, and other activities is very hard, and is one of the most difficult things I have ever done. However, it was worth every moment. Shocking, to me at least, is that there is no other book out on the state fossils—because each and every state fossil has something to say both about the science and the state it comes from.

Thirty-four states designate state fossils, four have state dinosaurs, four name fossils as state stones, and two recognize fossils as state gems. Although some of these fossils are common and well-known from many areas, such as the mammoth, the majority are either unique or very important to the state that has designated them. It would be very easy for each state to name *Tyrannosaurus rex* as its state fossil, but most every state has chosen to recognize a fossil that has true meaning, not one that is simply ostentatious or media-popular.

Very few people have likely heard of *Pertica*, the state fossil of Maine, or *Ecphora*, Maryland's state fossil. Almost as likely, few probably find these fossils, a Devonian plant and a Tertiary gastropod, too impressive. Certainly, ABC News wouldn't send a team out to cover the discovery of a new specimen, as happens with the large, more "showy" fossils. However, these fossils are not unimportant, but in fact say much about the states that designated them.

Pertica, an early vascular land plant, is only known from Maine. While most collectors would certainly rather have a complete mammoth than a slab of *Pertica* specimens, the fact remains that this plant has given scientists new insights into the early evolution of land plants. In addition, it is only found in one place: Maine. *Pertica* is a perfect example of a perfect state fossil.

Illinois' Tully Monster is another example of an ideal state fossil. I live in Illinois, and to the average collector here, there are few animals that inspire more awe than the Tully Monster. Found only in a handful of spots in Illinois, this Pennsylvanian invertebrate is a mystery, with science unable to decipher its ancestry or modern relatives. Illinois could have easily laid the title of state fossil on one of the popular mammoths or mastodons found in the northern part of the state, but instead the State Legislature chose the Tully Monster.

Indeed, most state fossils tell much about the states they represent. They tell about the ancient environment, the modern state of geological and paleontological science in that state, the popularity of certain fossil specimens, and even the goals and desires of the state government. Often, the legislature recognizes fossils that it sees as "ubiquitous and popular with collectors," or specimens that will provide "an educational resource to the students of the state."

More often than not, the state fossil of a certain state was either originally discovered there, is unique to the state, or has a well-documented history with collectors. Such is the case with *Astrodon*, the state dinosaur of Maryland, and *Hadrosaurus*, the state dinosaur of New Jersey. Both were originally found in those states, and both have since become important in the study of dinosaurs.

While most state fossils are unique to their state, some are not. Three states recognize a type of petrified wood as their state fossil, and three more as their state stone or gem. Three states also recognize a mammoth, and dinosaurs, of course, are very popular. However, these fossils are exceptions to the rule. The most common state fossil is one that says a great deal about its state.

Paleontology is such a cutting-edge science that it is tough to find a fossil site untouched by scientists or collectors. Therefore, it is likely that outstanding discoveries will continue to be made across the U. S. Perhaps, some of these will provide state fossils for the 16 states that do not currently recognize them.

Until then, the text you are about to read represents information on every state fossil, state dinosaur, and every fossil designated as a state gem or state stone. Also, at the end, is included a section on some recent entries that have been proposed as state fossils. Happy reading!

STATE FOSSILS

ALABAMA

BASILOSAURUS

Although dinosaurs and large land mammals get much publicity in popular paleontology books, one of the most fascinating groups of animals to have ever lived is the whales. Members of the Order Cetacea, whales can trace their ancestry back to a group of primitive terrestrial ungulates (hoofed mammals) known as artiodactyls. Much of the information we now know about the artiodactyl-whale relationship can be traced to two places: Pakistan and the Mississippi-Alabama area. In fact, primitive whales are such common fossils in Mississippi and Alabama that each state has adopted a whale as its official state fossil.

During the Eocene Period, some 45 million years ago, the lands that make up modern-day Alabama were under water. This aquatic environment was a haven for archaeocetes, or primitive whales. One of these archaeocetes was *Basilosaurus cetoides*, Alabama's official state fossil.

In order to understand the importance of *Basilosaurus*, we must travel back to well before this whale lived, to the Alabama of 63 million years ago. At this time, in the early Paleocene Epoch, artiodactyls were beginning to evolve in the Alabama-Mississippi-Louisiana area. However, for some reason still unknown to modern science, these small hoofed mammals took a strange evolutionary course. By the dawn of the Eocene, some of these ungulates had changed so drastically that they were poised to leave land and enter the water.

Of course, evolving from a terrestrial vegetarian to an aquatic carnivore required many adaptations for living in the water, including a backwards and upwards shift of the nostrils, the appearance of coverings for the nostrils, loss of the rear limbs, streamlining of the shape, change of the forelimbs into flippers, the addition of flukes for swimming, loss of most hair, the addition of an insulating layer of blubber, and the modification of senses for use in the water. The first animals to show these changes were the archaeocetes.

The archaeocetes, members of the Suborder Archaeoceti, possessed tiny heads and pointed snouts with differentiated teeth, much like those of most land mammals. In contrast, modern whales either lack teeth or have teeth that are virtually all identical in shape and size. Archaeocetes also had nostrils near the tip of the nose, another feature of land mammals, and lacked a blow hole. Some retained substantial hind limbs that would have been visible outside the animal's body. In the earliest archaeocetes, these limbs and the pelvis were attached to the vertebrae by a sacral joint. Over time, this feature was lost, and in modern whales only a vestige of the hind limb remains. This vestigial appendage is not attached to the rest of the skeleton.

By the late Eocene, some 45-35 million years ago, archaeocete whales had spread to many parts of the world, including the Gulf Coast area. *Basilosaurus* was one of the most common of these Alabama whales.

Basilosaurus itself resembled a sea serpent, mostly due to its tiny head, flippers, and narrow body. In life, it may have grown to lengths of 70-80 feet, with most specimens measuring in the 50-foot range. *Basilosaurus* is characterized by its reduced hind legs and pelvis and lack of rear

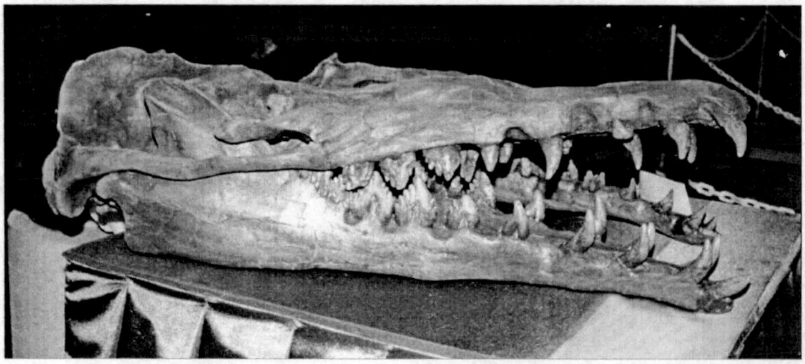

flippers, a definite reminder of its terrestrial origins. Its narrow, wedge-shaped, 5-foot-long jaws held 44 teeth, among them well-developed incisors, canines, premolars, and multi-rooted molars. These teeth were divided into two basic shapes, with cone-shaped teeth (likely used to hold prey) in the front of the jaw and triangular-shaped teeth (likely used for slicing) in the rear. The type and amount of wear on the teeth of *Basilosaurus* indicates that it probably ate squid and fish. Furthermore, these teeth were much like those of early placental mammals, including *Basilosaurus*' artiodactyl cousins.

The first known *Basilosaurus* material was discovered in 1832 by a wealthy Arkansas physician from the bluffs along the Ouachita River. This specimen, consisting of 28 vertebrae, was described in 1843 by local physician Dr. Richard Harlan, who thought he was describing the backbone of a large reptile. Because of this error, he named the animal *Basilosaurus cetoides*, which means "king of the lizards."

Two years later, in 1834, a complete skeleton, the first discovered, was found on a plantation in southwestern Alabama. Staff of the Philadelphia Academy of Natural Sciences, along with Sir Richard Owen, studied the skeleton and concluded that it was not a lizard, but a meat-eating member of the whale family. Because of this, Owen renamed it "Zeuglodon," meaning "yoked-tooth," and estimated it to be over 45 million years old. However, under the rules that govern the naming of species, the earliest name for a species has priority, even if it is inaccurate. Therefore, the name *Basilosaurus* is still used.

In the years immediately preceding and following the 1834 discovery, fossil remains of *Basilosaurus* were also discovered in Clarke, Choctaw, and Washington Counties. One of these discoveries, made in 1892 in Washington County, is now displayed at the Red Mountain Museum in Birmingham.

The frequent discovery of these prize specimens drew many scientists to Alabama. Unfortunately, it also drew many other people, among them promoter Albert Koch. Koch came to Alabama in 1845 to find rare fossils, and was appalled to hear that a New York paleontologist would be arriving shortly to collect "Zeuglodon" bones. Koch rushed to the site and secured the bones, barely beating the scientist to the treasure. Despite a shipwreck in which the bones were temporarily lost, Koch brought the fossils to New York City and exhibited them as a 114-foot-long sea serpent! This was quickly denounced by anatomists and paleontologists, but Koch's exhibit still drew large crowds.

Today, the "Basilosaur Belt" of parts of Alabama, Mississippi, and Louisiana continues to yield fine whale fossils. Most of these are discovered in sediments that make up the Jackson Group, including the Moodys Branch Formation and the Yazoo Clay. In fact, these fossils are so common that local residents have been known to use vertebrae of *Basilosaurus* for fireplaces and blocks to support cabins.

In 1984 *Basilosaurus* was officially declared the state fossil of Alabama with the passage of Act No. 84-66. Today, good specimens of this whale can be found in the Red Mountain Museum in Birmingham, the National Museum of Natural History in Washington, the University of Southern Mississippi in Hattiesburg, the Mississippi Office of Geology in Jackson, and the Mississippi Museum of Natural Science in Jackson. *Basilosaurus* fossils have also been discovered in Australia and Egypt, and remain as some of the most important pieces of evidence in understanding the evolution of early whales.

FOR FURTHER INFORMATION:

Brusatte, S., 2001. Kids' Corner-*Basilosaurus*. Fossil News: Journal of Avocational Paleontology, v. 7, no. 6, p. 11.

Dockery, D. T., and Johnston, J. E., 1986. Excavation of an archaeocete whale, *Basilosaurus cetoides* (Owen), from Madison, Mississippi. Mississippi Geology, vol. 6, no. 3, pp. 1-10.

Gingerich, P. D., Smith B. H., and Simons, E. L., 1990. Hind limbs of Eocene *Basilosaurus*: evidence of feet in whales. Science, vol. 249, pp. 154-157.

Gingerich, P. D., Wells, N. A., Russell, D. E. and Ibrahim Shah, S. M., 1983. Origin of Whales in Epicontinental Remnant Seas. Science. vol. 220, pp. 403-406.

Heinrich, P. V. Primitive Eocene Whales. http://www.intersurf.com/~heinrich/Basilosaurus1.html

Kellog, A. R., 1936. A review of the Archaeoceti. Carnegie Institution of Washington Publication no. 482, 336 pp.

Thurmond, J. T., and Jones, D. E., 1981. Fossil Vertebrates of Alabama. The University of Alabama Press, University, Alabama.

ALASKA

MAMMUTHUS PRIMIGENIUS

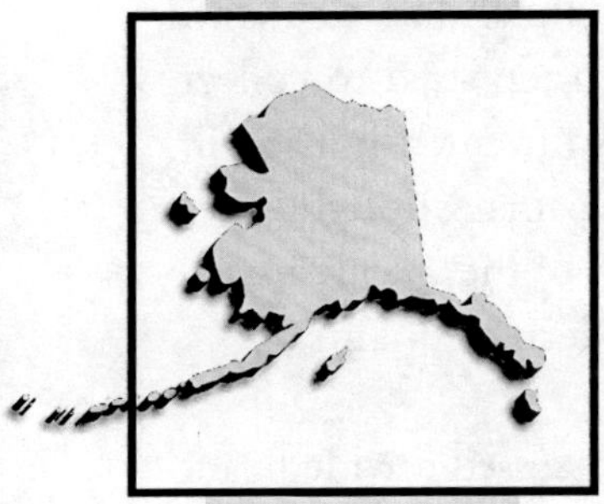

If a person thinks of the climate of North America during the most recent Ice Age, the first thought that pops into his or her head is likely of a snow-covered, freezing environment chock full of large mammals. While this may seem exotic or strange to people living in the Sunbelt, it isn't far-fetched for the residents of Alaska, often dubbed by scientists "the Ice Age state." It is fitting, then, that Alaska has designated one of the best-known Ice Age animals as its state fossil, the woolly mammoth.

The most recent Ice Age, perhaps one of the best-known geological Epochs, was a time of strange climate and an equally strange fauna. Lasting from about 1.6 million to 10,000 years ago, this time, known as the Pleistocene, saw the evolution of large mammals, the continued development of humans, and the reign of the woolly mammoths.

Woolly mammoths are known to science as *Mammuthus primigenius*. These large animals were members of the Family Elephantidae, a group of mammals that includes the modern elephants and the extinct gomphotheres. Mammoths likely first evolved in Europe and Asia, and later made their way into North America via the Bering Land Bridge. All woolly mammoths are immediately recognized in paintings and reconstructions by their signature coat of long, dense, dark brownish hair. This hair, which is the etymology of the mammoth's moniker, has been found frozen in Siberia, and is therefore well-studied. It grew to lengths of several inches and was used as an insulatory structure.

Woolly mammoths are also recognized by their pair of long, curved tusks. These tusks, which grew to lengths of several feet, were probably used to gather vegetation and to wrestle with other mammoths. Based on a study of modern elephants, including the woolly mammoth's closest relative, the Indian elephant, these tusks are formidable food-gathering tools, and are often used to collect vegetation that lies low to the ground. The tusks are also frequently used in intraspecies wrestling matches, competitions between two males fought over territory or mates. The same uses are inferred for the tusks of the woolly mammoth, although they were much longer than those of modern elephants.

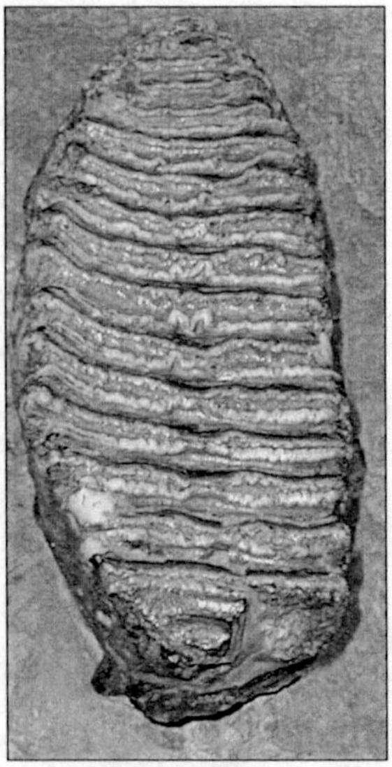

Woolly mammoths also possessed a wide head, a sloping back, a fatty hump, large ears, and a proboscis (trunk). The average woolly mammoth was about 3.5 meters (11-12 feet) long, 3 meters (9.5 feet) tall at the shoulders, and weighed about 3 tons. Woolly mammoths were grazers on grasses, sedges, and flowering plants. As a general rule, woolly mammoths preferred steppe-like grassland habitats, indicating that parts of Alaska were more temperate during the Ice Age than they are now. The best scientific estimates are that a mammoth's individual lifespan was around 70-80 years. During this time, the mammoth would go through six sets of flat molar teeth. The biggest of these teeth, which are typically found as fossils, were about the size of a large shoe box and had a flat, ridged surface adapted for grinding up tough plant material.

While woolly mammoths had a slew of special adaptations to a successful grazing lifestyle, the fact remains that they became extinct around 10,000 years ago, at the end of the Pleistocene. Why? The question of their extinction has puzzled paleontologists for decades. Originally, it was thought that the inability of the mammoth to adapt to a new, warmer climate led the species to die out. Ice Age winters were colder but also drier than now. Many scientists believe that changing vegetation and increased snowfall—through which the mammoths may have had difficulty digging for winter forage—did them in. However, in recent years other scientists have dropped this traditional view, and have instead pointed to humans.

Fossil mammoth bones, especially those discovered in Nebraska and New Mexico, have shown evidence of human interaction. At a few isolated sites, spear tips and other evidence of human hunting have been found alongside mammoth and other Pleistocene mammal bones, suggesting that these individuals were killed by early human hunters. The fact that humans are known to have killed mammoths has led some scientists to hypothesize that human predation led to the extinction of mammoths as a group. According to a recent study by University of California at Santa Barbara biologist John Alroy, it is more than plausible that human hunters entirely wiped out the mammoth and other large Ice Age mammals. However, American Museum of Natural History paleontologist Ross MacPhee believes that humans didn't cause the extinction of the mammoth by overhunting, but rather by spreading a new, deadly disease. This hypothesis has also gained much weight in recent years. Regardless of exactly how mammoths died out, it is very possible that humans did play a hand in one of the most devastating extinctions of all time.

Because of the woolly mammoth's recent (in geological terms) extinction, it was thought during much of the 19th century that mammoths

may still have survived in the extreme Northern Hemisphere, including Alaska. In 1899, for instance, a magazine printed a story claiming that the last mammoth was killed in the Yukon Territory, Canada, in 1891. It was a hoax, but many believed it. Those who believed the mammoth still existed pointed to the discovery of mammoth tusk ivory throughout the Arctic territories, and mammoth paintings in the caves of Europe. While the mammoth paintings were actually the leavings of Ice Age hunters, the ivory of mammoths can still be found today, several thousand years after the mammoth became extinct. Why? Because of the occasional thawing of permafrost in many regions of Alaska, well-preserved mammoth remains turn up frequently, including tusks. These large tusks are often found during construction projects, when they are released by erosion from rivers, and also occasionally turn up during gold-prospecting expeditions.

One of the earliest reports of mammoth remains in Alaska came in 1836, when explorers announced the discovery of strange fossil bones found on the Pribilof Islands. At least five additional major mammoth discoveries have been made on these islands since the explorers first found bone. Later, in 1897, mammoth bones were found in a volcanic cave on the island of St. Paul, leading paleontologists and explorers to comb the entire state, looking for remains. One of the best-preserved mammoth specimens was found in 1952, when a 9.75-inch-long mammoth tooth was found in the aforementioned cave on St. Paul. Although teeth are common, this specimen was special because it was well-preserved and found in an unlikely place—a spot where scientists had been looking for several years. Mammoth remains continue to turn up in Alaska, including along the Kuskokwim

River and near Cape Deceit where prized specimens have been recently found, and in the nearby Yukon. Mammoth ivory is also regularly found, and is often polished, carved, and sold to tourists in Alaskan souvenir shops.

Because of its popularity with the public and the abundance of fossils found in Alaska, the woolly mammoth was named the official Alaska state fossil in 1986, with the passage of Act 44.09.120. While the Ice Age may seem remote and dreamlike to many, this cold and glacier-filled age reminds Alaskans of the current environment of their state. It is perfect, then, that the best-known Ice Age animal is Alaska's official state fossil.

FOR FURTHER INFORMATION:

Bocherens, H., Fizet, M., Mariotti, A., Gangloff, R.A., and Burns, J.A., 1994. Contribution of Isotopic Biogeochemistry (C,N,O) to the Paleoecology of Mammoths *(Mammuthus primigenius).* Historical biology, vol. 7, p. 187-202.

Brusatte, S., 2002. Kids' Corner-mammoth. Fossil News: Journal of Avocational Paleontology, v. 8, no. 9, p.18-19.

Dawson, M.R., 1967. Fossil History of Recent Mammals, in: Recent Mammals of the World: A Synopsis of Families, S. Anderson and J.K. Jones, Jr. (eds)., Ronald Press, New York, 453 p.

Keddie, G., The Mammoth Story. Royal British Columbian Museum, http://rbcm1.rbcm.gov.bc.ca/hhistory/mammoth/mammothstory.html

Kurten, B., and Anderson, E., 1980. Pleistocene Mammals of North America. Columbia University Press, New York, 244 p.

Maglio, V.J., 1973. Origin and Evolution of the Elephantidae. Transaction of the American Philosophical Society, vol. 63, p. 149.

Nash, J.M., 2001. Who Killed Woolly. Time Magazine, vol. 157, no. 24, p. 64-65.

ARIZONA

ARAUCARIOXYLON ARIZONICUM

Fossils are a natural resource treasured by many, but often protected by few. A notable exception to this preservation paradox is the famous Petrified Forest National Park, a large section of the northeastern Arizona desert preserved from the ravages of developers and commercial fossil collectors. The highlight of this National Park, one of the most popular tourist sites in Arizona, is the object of its name: petrified wood. This petrified wood is so popular that it has been named Arizona's official state fossil.

Petrified wood, representing several types of trees and periods of time, has been found in nearly every Arizona county. The majority of this wood is found in the Chinle Formation, a Late Triassic rock unit that preserves both abundant plant and animal fossils. During the time this formation was laid down, about 225-208 million years ago, Arizona was composed of a lowland basin, bounded to the north and west by a receding sea and to the south by a chain of high-peaked mountains. Streams leading from these mountains provided the valley with sediment, which was both a blessing and a curse for the abundant plant life which called the valley home. On the one hand, these streams provided fertile soil in which to grow. On the other, they often flooded the valley with too much water and sediment, burying entire forests of trees and populations of animals. These occasional floods have provided paleontologists with the beautiful array and variety of petrified wood found in Arizona today.

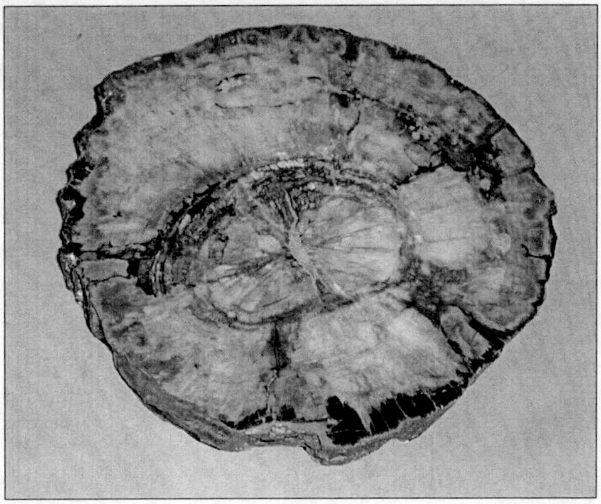

The most plentiful type of petrified wood found in Arizona is that derived from conifers. Of these conifers, *Araucarioxylon arizonicum* is the most common, making up some 90% of the petrified wood discovered in Petrified Forest National Park. In life, these gigantic trees would have grown to upwards of 150-200 feet, and possessed a trunk diameter of 4-5 feet. Many of these trees grew together in large stands, providing the basin with a thick vegetative canopy cover.

Although the phylogenetic relationships of *Araucarioxylon* are not well-understood, the wood of this tree superficially resembles that of the Family Araucariaceae, better known as the araucarians. Today, the araucarians survive in the southern hemisphere as such rare and strange species as South America's monkey-puzzle tree and the common Norfolk Island Pine, often grown as a house plant. While *Araucarioxylon* is not necessarily related to the araucarians *per se*, its resemblance to these modern trees led to its scientific name, which basically means "araucarian wood."

A problem in paleobotany is that for a given plant, the stem, leaves, and cones or flowers are rarely found fossilzed together and attached. This has led to the use of "form genera" to specify each of those components, when it is unknown in reality whether or not they actually belonged to the same biological species. A form genus may contain similar parts from possibly unrelated plants, as in the case of *Araucarioxylon* wood (which perhaps represents a variety of coniferous trees); or alternatively, two form genera may be used for two separate parts of the same plant—say, a leaf and a cone. It makes it difficult for the non-specialist to keep track of things, but at the current state of paleobotanical knowledge, it's the best that can be done.

Alongside *Araucarioxylon* grew distantly-related conifers such as *Woodworthia* and *Schilderia*. *Woodworthia* was characterized by a straight, unbranched trunk and likely lived in wet areas. *Schilderia* was much smaller, and possessed a trunk with a diameter of two feet. These two conifers are represented in Petrified Forest National Park, but are better known from other areas of Arizona.

Ancient horsetails (scouring rushes) grew closer to permanent water sources. These plants are similar to the horsetails growing today, with straight and slender hollow stems, but were much larger than the modern species of *Equisetum*. *Neocalamites* and *Equisetites*, two Triassic horsetails that grew up to 30 feet tall and had trunks 16 inches in diameter, were both identified in Petrified Forest National Park, but are also better known from other areas of the state. Two types of cycads, gymnospermous plants that resembled large pineapples with a crown of palmlike leaves, have also been found in the park. *Lyssoxylon*, the larger of the two species, grew to heights of three to four feet, while *Charmorgia* was much shorter, and only reached heights of one foot. Ferns represent the most abundant group of leaf fossils found in the park. Many of these ferns grew to heights of 10 feet, making them much larger than most modern ferns, and on par with the tree ferns of New Zealand and Australia.

The first specimens of petrified wood in Arizona were discovered by various Native American tribes, who often concocted elaborate stories and myths to explain their existence. For example, the Piutes believed that the petrified logs were spent arrow shafts discarded by their thunder god, Shinauav. Other tribes even built dwellings out of the logs, and many used the hard minerals as the raw material for fashioning arrowheads and other tools. However, the first discovery of Arizona's petrified wood by American settlers is much better documented. This discovery can be traced back to a decorated Civil War general, William Tecumseh Sherman, and his drive to secure prized fossil specimens for the Smithsonian Institution.

In the autumn of 1878, while exploring the Arizona Territory, Sherman came across what he later described as "extraordinary specimens of petrification." As a regent of the new Smithsonian Institution, Sherman wrote a letter to his cohort, Colonel Peter Swaine, telling him to collect some interesting samples for the museum. In the spring of 1879, Swaine dispatched Lieutenant J.F.C. Hegewald of the 15th Infantry with a sergeant, 12 men, and two heavy mule-pulled wagons to the area that Sherman had described a year earlier. At Navajo Springs, Arizona, Hegewald reported meeting with a group of Navajo tribesmen, who insisted that the logs represented the bones of giants, slain by their forefathers many centuries ago. However, the Navajos led Hegewald to thousands of petrified specimens, two of which the Lieutenant managed to collect and transport back to the Smithsonian, where they still stand today.

Sadly, soon after Hegewald transported the two specimens back to Washington, news began to spread of the "petrified forests." Largely because of descriptions in European and Asian magazines, by 1885 the trees had become very popular with foreigners. It was even reported that one Russian dealer had paid $500 for a large log, which he planned to slice up and make into table tops. Soon after, Drake & Co. of Sioux Falls, South Dakota began buying specimens to make into mantels, clock cases, and paperweights, many of which were sold to Tiffany & Co. In 1899 a U.S. Geological Survey report declared that if something wasn't done about the situation soon, the ancient forest would disappear. Seven years later, in 1906, President Theodore Roosevelt designated the area a National Monument. However, it would take until 1962 before President John F. Kennedy made the area a National Park. Today, this 93,492-acre area, straddling Interstate 40 east of Holbrook, is known as Petrified Forest National Park.

Today, the park is one of Arizona's top tourist attractions. Visitors are drawn by the plethora of ancient trees. The most widely-accepted hypothesis regarding the circumstances of their preservation is that the conifers grew in a warm climate with seasonal monsoon rains. This seemingly peaceful environment was interrupted by a sudden volcanic eruption in the highlands to the southeast, which overloaded the banks of nearby rivers and streams with ash and debris. This debris caused rivers to overflow, which led to a toppling of the forests by powerful floodwaters. Ash quickly covered the downed trees, burying them at least 30-40 feet under the surface. Shut off from air and bacteria, the organic components of the trees were gradually replaced by minerals. Millions of years later, these crystalline minerals leave us with vividly-colored fossil wood.

To preserve this valuable resource, wood cannot be collected at Petrified Forest National Park, and any illegal collecting is subject to a large fine. However, good specimens can be collected at several other sites across the state, and (with permission) on private land outside of the park boundaries. For those more interested in seeing specimens in museums, petrified wood from Arizona can be seen both at the Park and in the Smithsonian, and is stored in the collections of several of the state's best museums.

Because of its popularity with collectors, tourists, and paleontologists, petrified wood was named the official Arizona state fossil in 1988. While Arizona has a diverse fossil record, including a glut of footprints, dinosaur remains, and early reptile bones, it is the less sensational petrified wood which has garnered the most awe from both the state residents and the state's legislature.

FOR FURTHER INFORMATION:

Ash, S.R., and Creber, G.T., 2000. The Late Triassic *Araucarioxylon arizonicum* trees of the Petrified Forest National Park, USA. Paleontology, vol. 43, part 1, p. 15-29.

Brusatte, S., 2001. Kids' Corner-Petrified Wood. Fossil News: Journal of Avocational Paleontology, vol. 7, no. 1, p. 10-11.

Conover, A., 1997. The Object at Hand. Smithsonian Magazine, no. 6.

Long, R., and Houk, R., 1988. Dawn of the Dinosaurs: The Triassic in the Petrified Forest. Petrified Forest National Park, 96 p.

Miller, A.C., Trees of the Petrified Forest National Park, http://www.geo.arizona.edu/geos256/azgeology/pwood/trees.html

NPS, Petrified Forest National Park, http://www.nps.gov/pefo/

White Mountains Online, Petrified Forest National Park, http://www.wmonline.com/attract/pforest/pforest.htm

CALIFORNIA

SMILODON CALIFORNICUS

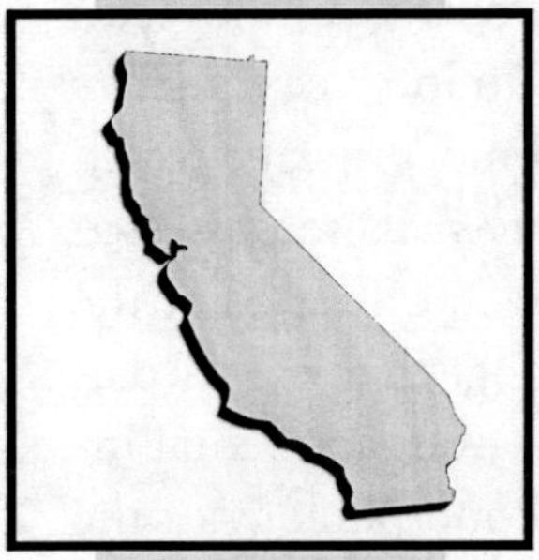

Few state fossils are as immediately recognizable as California's. While dinosaurs are popular state fossils in the Rocky Mountain states and invertebrates receive most of the attention out east, it is the Ice Age fauna which garners the most study in California. Based on the study of a handful of prized "super sites," scientists have learned a great deal about the morphology, lifestyle, and evolutionary relationships of Pleistocene animals. Among the most popular, well known, and best studied of these Ice Age animals is the saber-toothed cat *Smilodon californicus*, California's state fossil.

Smilodon, one of the best-known of the saber-toothed cats, first appeared about 1.6 million years ago. It lived during the last great Ice Age, a time in which the Earth was overrun by glaciers. These glaciers led to a drastic cooling of the Earth's temperature and the evolution of a strange megafauna that included such interesting animals as the giant ground sloths, mastodons, mammoths, glyptodonts, and, of course, the saber-toothed cats. Although its world was much colder than our modern climate, *Smilodon* lived in a habitat that is not unlike what much of California is today: shrubby grasslands and pine forests.

In life, *Smilodon* was the largest of the saber-toothed cats, a morphotype which evolved three times in the Family Felidae. It was a fierce predator, measuring about 1.2-1.5 meters (4-5 feet) in length and about 1 meter (3 feet) tall. Measurements based on fossils estimate its weight to have been about 200 kilograms (440 pounds). These dimensions would have made it a bit smaller than a modern-day lion, but much heavier. As a general rule, *Smilodon* possessed relatively short legs and a short, bobbed tail. Its front legs were powerful, and its body was adapted for springing onto prey. However, it wasn't a very fast runner.

The skull of *Smilodon* is immediately recognized by two large canine teeth. These saber-teeth were serrated, oval in cross section, and up to 18 centimeters (7 inches) in length. These canines contrast with those of modern cats, which have noticeably smaller teeth. Unbeknownst to some, *Smilodon* also possessed a pair of lower canines, which were much reduced in size and located at the front end of the lower jaws. There was also a mid-sized flange that was used as a guard for the upper canines, serving to protect the skull from accidental injury. As a result of the development of these canines, the morphology of the *Smilodon* skull was much different than that of modern lions. Nasal openings were reduced from the position in typical cat skulls, and prominent mastoid processes were evolved for the attachment of strong muscles used to pull the head down. These muscles helped *Smilodon* open its jaws to an angle of about 120° and enabled the animal to swing its head at almost a right angle when it attacked.

It is obvious that *Smilodon* was a successful predator. However, the body and limbs of this animal have led scientists to believe that its hunting strategy was much different than that of modern lions and wild cats. While modern lions usually chase their prey over a large distance, causing the quarry to succumb to exhaustion, *Smilodon* was likely an ambush predator. Its lower legs (tibiae and fibulae) were short, signs of a slower animal. This probably meant that the cat ambushed and chased slower animals, such as mammoths and mastodons, rather than fleet-footed ungulates like cervids (deer family). This is only speculation, but *Smilodon* may have gripped its prey with its strong front limbs and used its teeth to stab a vulnerable spot, like the stomach or jugular. The prey would then bleed to death, and *Smilodon* would feed on its quarry.

Another hard-to-prove notion is the social relationships of *Smilodon*. Many *Smilodon* fossils show evidence of serious crushing and fracture injuries, yet frequently these bones show signs of healing and regrowth. How could saber-tooths survive these life-threatening injuries? It has been hypothesized that these cats lived in packs and that injured individuals were brought food by other saber-tooths. Some *Smilodon* remains show bite marks that match up to the jaws of other *Smilodon* individuals, possibly marks of intraspecies fights.

Smilodon was a member of the Family Felidae, a group of carnivorous mammals that includes modern lions, tigers, and smaller cats. The saber-tooth morphology has evolved at least three times within the Felidae, its expression in *Smilodon* being the most pronounced. While these lengthened canines were certainly a successful hunting adaptation, *Smilodon* and its fossil relatives no longer remain. Why? The extinction of the saber-tooth may have been related to the decrease in large animals upon which it preyed. Whether due to habitat disruption or human hunting, mammoths and mastodons disappeared at the close of the Pleistocene, animals which were likely favored prey of *Smilodon.* Or, possibly a disease wiped out the megafauna. The only surety is that *Smilodon* disappeared about 11,000 years ago, at the end of the last Ice Age.

Over 10,000 years later, remains of *Smilodon* frequently turn up in the middle of downtown Los Angeles. These remains come from the world-famous Rancho La Brea Tar Pits. These pits, recognized as one of the best fossil sites in the United States, can trace their origin to 40,000 years ago, when the area was covered by pools of heavy, sticky, black tar. This sticky tar was the remnant of crude oil which had seeped through to the surface. Over a period of several thousand years the

light fraction of the oil evaporated, leaving behind the heavy tar in sticky pools. During the last Ice Age, these pools were death traps for large animals.

The La Brea pits were first discovered by Native Americans, who for thousands of years used the tar as a waterproofing material for baskets and canoes. After the arrival of Europeans, the tar from the pits was mined and used for roofing by the inhabitants of the town of Pueblo de Nuestra Señora la Reina de Los Angeles, now simply known as Los Angeles. The citizens of Los Angeles frequently discovered bones in the tar, but dismissed them as belonging to unlucky cattle. It was not until 1901 that the first scientific excavation of the pits was carried out. Scientists from the University of California at Berkeley, most notably Professor John C. Merriam and his students, were among the first academic researchers to excavate and describe the La Brea fossils. Many years after their initial excavations, in 1932, Merriam and some students published the first monograph on *Smilodon*, based on fossils found at La Brea.

Since Merriam's initial excavations, some 160,000 specimens of *Smilodon* have been discovered at the pits, coming from an estimated 2,000 individuals. This makes *Smilodon* the second most common animal at the pits. In nature, herbivores generally outnumber carnivores by about 10:1, yet the La Brea fossil assemblage is preferentially skewed towards carnivores. Why? If a pack of carnivorous mammals were to chase a lone prey animal into a tar pit, all of them would become trapped. Over time, this would create an abundance of carnivores. In all, some 59 species of mammals have been found at the site. Many of these fossils are exhibited today in the George C. Page Museum of La Brea Discoveries, which opened in 1977. *Smilodon* remains can also be seen at the University of California at Berkeley's Paleontology Museum and at the University of California at Davis, along with a large number of other museums worldwide.

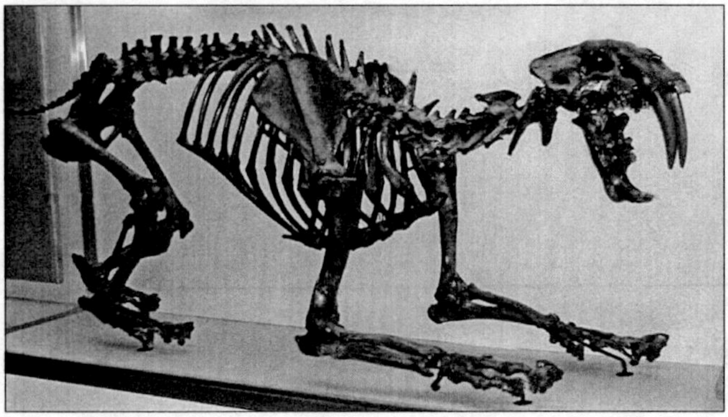

Because of its abundance and scientific importance, the "saber-tooth cat" was named the official California state fossil in January of 1974. This designation resulted from a campaign sponsored by the magazine *California Geology*, which called for the proposition of various candidates for a state fossil. While trilobites, snails, molluscs, and a sea lion were nominated, it was the saber-tooth that received the honor with the passage of Government Code Section 420-429.5.

Few state fossils are as easily recognized as California's, and even fewer state fossils are as abundantly known. Because of the amazing discoveries at Rancho La Brea, the lifestyle, morphology, and evolution of *Smilodon* and its relatives are well-known. And, because of these unusual tar pits, millions of California residents are aware of their state's amazing Ice Age past.

FOR FURTHER INFORMATION:

Akersten, W.A., 1985. Canine Function in *Smilodon* (Mammalia; Felidae; Machairodontinae). Contributions in Science, Natural History Museum of Los Angeles County, 356: 1-22.

Brusatte, S., 2000. Kids' Corner-*Smilodon*. Fossil News: Journal of Avocational Paleontology, vol. 6, no. 6, p. 15.

Harris, J.M., and Jefferson, G.T. (eds.), 1985. Rancho La Brea: The Treasures of the Tar Pits. National History Museum of Los Angeles County.

Turner, A., 1997. The Big Cats and Their Fossil Relatives. Columbia University Press, New York.

Discovery Channel, Sabre-Tooth, http://dsc.discovery.com/stories/nature/sabretooth/sabretooth.html

George C. Page Museum of La Brea Discoveries, http://www.tarpits.org

Natural History Museum of Los Angeles County, Cats! Wild to Mild, http://www.lam.mus.ca.us/cats/encyclo/smilodon/

University of California at Berkeley, Sabre-Toothed Cats,
http://www.ucmp.berkeley.edu/mammal/carnivora/sabretooth.html

COLORADO

STEGOSAURUS STENOPS

Although its land is rugged and its terrain rocky, Colorado is one of the most popular vacation destinations in the United States because of its beauty. Dinosaur hunters and paleontologists come to Colorado, too, but their mission isn't only to relax and enjoy the fresh mountain air. Much of Colorado is underlain by rocks belonging to the Morrison Formation, a Jurassic unit which contains the well-preserved remains of countless species of dinosaurs. One of these dinosaurs is *Stegosaurus stenops*, the familiar plated herbivore popular with school children of all ages. This popularity, along with its scientific importance, led the Colorado State Legislature to declare *Stegosaurus stenops* the official state fossil.

The Morrison Formation, named for the tiny Colorado town of Morrison, west of Denver, stretches over a vast area of mountains, forests, grasslands, and flatlands. However, during the Jurassic, the land that is now Colorado was was a vast coastal floodplain, much lower in elevation than it is today, and was dotted with rivers, lakes, and streams—environments perfect for fossilization. The result is the plethora of dinosaur fossils found in Colorado today, among them *Stegosaurus*.

Stegosaurus was an amazing and intriguing herbivore. This dinosaur, distinguished by its plates and spikes, grew to lengths of 8-9 meters (26-30 feet), and to heights of about 2.5-3 meters (8-10 feet). The average *Stegosaurus* may have weighed up to 3,100 kilograms (6,800 pounds), but its brain was only the size of a walnut, weighing roughly 2.5-3 ounces. The skull of *Stegosaurus* was long and narrow, and was characterized by a toothless beak and small cheek teeth. The back legs of *Stegosaurus* were twice as long as its front legs, which meant that the individual was an obligatory quadruped. However, some paleontologists believe that *Stegosaurus* may have reared up on its large hind legs to reach higher vegetation.

The most distinctive feature of *Stegosaurus* was its plates. These triangular plates, which varied in number, were embedded in the skin along the backbone of *Stegosaurus*. As a rule, the plates began at the base of *Stegosaurus*' head, and were quite small. They became increasingly larger and larger at the middle of the back, and then smaller again down the tail, stopping halfway down its length. The placement of these plates, unlike their size, has been a controversial issue for over a hundred years. When the first specimen of *Stegosaurus* was discovered, it was thought that the plates were arranged in a single row down the back. However, based on a 1992 find in Colorado, Denver paleontologist Ken Carpenter believes that the plates ran along the *Stegosaurus*' back and tail in two rows, alternating in alignment.

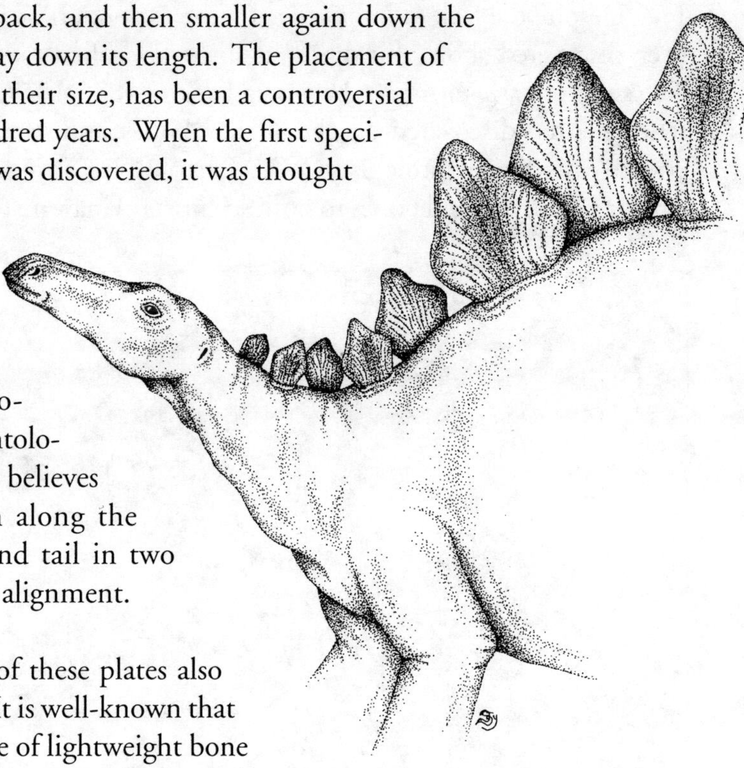

The function of these plates also remains a mystery. It is well-known that the plates were made of lightweight bone and were filled with tube-like tunnels. These tunnels likely mean that the plates were nourished by a mass of blood vessels, indicating that they may have been used to regulate the dinosaur's body temperature. There is also the possibility that they were used for display purposes, or possibly for protection. *Stegosaurus* also possessed a set of spikes at the end of its tail, which grew to lengths of four feet. These spikes, which numbered four in *Stegosaurus stenops*, were no doubt used for protection against predators.

Stegosaurus was a successful herbivore, and its lineage extended throughout much of the Jurassic. However, the curtain fell on *Stegosaurus* and many of its contemporaries at the end of the Jurassic. Why did this extinction occur? Many scientists believe that the evolution of new plants interrupted the feeding habits of *Stegosaurus*, although this hypothesis is unproven and the question may never be resolved.

Despite the lack of knowledge on the extinction of *Stegosaurus*, much has been discovered about this dinosaur since it was first found in the late 1800s. The story of *Stegosaurus* can be traced back to 1876-1877, when two Colorado schoolmasters discovered a cache of giant bones, and sent word of their discoveries to eastern paleontologists O.C. Marsh and E.D. Cope. One schoolmaster, Arthur Lakes, shipped bones to both scientists. Unaware that Cope had already

been sent some of the specimens, Marsh paid Lakes $100 to keep the discovery a secret. When Lakes received word that Cope was preparing to describe some of the fossils, he sent a note to the professor asking him to return the bones to Marsh. This outraged Cope, and ignited the infamous "bone wars."

This rivalry led to the discovery of the first *Stegosaurus* fossil, a specimen discovered by Marsh collector M.P. Felch during the spring of 1877. It didn't take Marsh long to name the specimen, as he christened it *Stegosaurus*, meaning "roofed plated reptile," later that year. Today this specimen is housed at the Smithsonian Institution in Washington, D.C. Marsh continued excavations near the site of the first *Stegosaurus* discovery, and during the years 1882-1888 he discovered a multitude of dinosaur fossils, including additional *Stegosaurus* specimens.

When Marsh was busy naming and excavating *Stegosaurus*, the second Colorado schoolmaster was beginning to aid his arch-rival, Cope. O.W. Lucas, the second of the teachers, had found and reported gigantic bones at Garden Park, near Cañon City. He later sent samples to Cope, who dispatched collectors to the site. Over the next several years these collectors discovered well-preserved specimens of *Allosaurus, Diplodocus,* and *Stegosaurus*. However, after Cope abandoned the site in the late 1880s, the Garden Park area fell into obscurity. Nearly a century later, in 1970, the area was designated a National Natural Landmark. In the 1980s a group of amateur paleontologists founded the Garden Park Paleontological Society, which pushed for increased protection of the area's fossil sites. Their efforts were rewarded when Garden Park was designated a Research Natural Area by the Bureau of Land Management in 1990. With their preservation efforts behind them, the Society today operates a small museum, the Dinosaur Depot, in Cañon City.

However, possibly the most important specimen ever discovered in the Garden Park area was found over 100 years after the departure of Cope. Nicknamed "Spike," this fossil represents the most complete *Stegosaurus* individual yet found. The specimen was discovered by paleontologists Bryan Small, Tim Seeber, and Ken Carpenter in 1992, and contained a nearly-intact skull and a rarely-preserved toe bone. Its plates, discovered in alternating rows, also helped resolve the century-old debate over the proper placement of these enigmatic structures. With the help of the Bureau of Land Management, the specimen was excavated and today resides at the Denver Museum of Nature and Science. Similar specimens can also be viewed at other museums across the country, including the American Museum of Natural History in New York and the Carnegie Museum in Pittsburgh.

With all the fanfare surrounding the discovery of *Stegosaurus*, its scientific importance is often lost. However, the State Legislature of Colorado paid homage to both the popularity and importance of this herbivore when it designated it the official state fossil. Credit for this designation goes to a fourth-grade class at McElwain Elementary School, which pushed for *Stegosaurus*' designation in 1980. The class researched the appropriate procedure for proposing and passing a legislative bill, and with the help of local paleontologists, the class lobbied for State Senate approval. Their campaign took the class across the entire state, where they presented informative exhibits on *Stegosaurus* at schools and community events. Their hard work was rewarded on April 28, 1982, when Governor Richard D. Lamm issued an executive order establishing *Stegosaurus* as the official Colorado state fossil. The State Legislature passed a similar bill in 1991, making it official.

Although *Stegosaurus* lived some 150 million years ago, this simple herbivore ignited a feud between respected scientists, caused debate in paleontology circles, and captivated the hearts and minds of a grade school class. It also starred in a major motion picture, *Jurassic Park*. Colorado is considered one of the richest paleontological states in the United States, and paleontologists from around the globe flock to the Centennial State to learn more about the world of dinosaurs during the Jurassic. Obviously, then, *Stegosaurus stenops* is an ideal choice for Colorado's state fossil.

FOR FURTHER INFORMATION:

Bakker, R.T., 1986. The Dinosaur Heresies: New Theories Unlocking the Mystery of the Dinosaurs and Their Extinction. William Morrow Publishers, New York.

Brusatte, S., 2001. Kids' Corner-*Stegosaurus*. Fossil News: Journal of Avocational Paleontology, v. 7, no. 11, p.18.

Colbert, E.H., 1968. The Great Dinosaur Hunters and Their Discoveries. E.P. Dutton & Company, New York.

Lessem, D., and Glut, D.F., 1993. The Dinosaur Society Dinosaur Encyclopedia. Random House, New York, 533 p.

Marsh, O.C., 1896. The Dinosaurs of North America. Sixteenth Annual Report of the United States Geological Survey, p. 133-244.

Olshevsky, G., 1995. Two Stegosaurs from One. Dino Doings.

Osborn, H.F., 1931. Cope: Master Naturalist. Princeton University Press, Princeton, New Jersey, 740 p.

Ostrom, J.H., and McIntosh, J.S., 1966. Marsh's Dinosaurs: The Collections from Como Bluff. Yale University Press, New Haven, 388 p.

CONNECTICUT

EUBRONTES GIGANTEUS

Today New England is a picturesque land, covered by rolling hills, dense pine and birch forests, and historical sites galore. But, 200 million years ago, during the Early Jurassic, it sported a much different landscape. The fauna consisted of phytosaurs, crocodiles, small reptiles, and a variety of dinosaurs, many of which left footprints that are abundantly found in Connecticut today. One of these footprints, the ichnospecies *Eubrontes giganteus*, has been designated the official state fossil of Connecticut.

The origins of *Eubrontes* can be traced back 200 million years. During the Early Jurassic, the Atlantic Ocean was just beginning to form as the supercontinent of Laurasia broke in two, sending present-day North America and Europe in opposite directions. Along the break, known as the Eastern Border Fault, a giant rift valley formed, comparable to those seen in modern East Africa. Powerful earthquakes dropped the valley floor down several hundred feet. Swift-flowing rivers from the mountains to the west gradually deposited over 16,000 feet of sediment into the rift, as the accumulating sediment depressed the original valley floor further. This sedimentary deposition created large, open mudflats, which were home to countless species of dinosaurs. Sheets of particles of the mineral mica, left by receding floodwaters, covered these mudflats periodically, preventing the mud from sticking to the feet of the dinosaurs while they walked. The result: thousands of well-preserved dinosaur footprints. Over time, additional layers of mud covered these footprints and turned them to stone, creating a plethora of footprint fossils, many of which are found in Connecticut today.

Eubrontes is one of the most common of these fossil footprints. This print belonged to a mid-sized, bipedal, predatory dinosaur. A typical *Eubrontes* track ranges from 40-50 centimeters (16-20 inches) long, and has three long toes, often with evidence of a heel. In addition, prints are usually spaced anywhere from 3.5 to 4.5 feet apart. Based on these measurements, the *Eubrontes* trackmaker must have been over one meter high at the hips and 5-6 meters long. One possible culprit is *Dilophosaurus*, the familiar crested theropod dinosaur from *Jurassic Park*. *Dilophosaurus* is known from the Early Jurassic, and reached a size comparable to what would be expected of the *Eubrontes* trackmaker. However, nothing short of a preserved *Dilophosaurus* skeleton fossilized in a *Eubrontes* track will conclusively pinpoint this Jurassic theropod.

Regardless of its questionable identity, much about the lifestyle of the *Eubrontes* trackmaker is known. For example, it has been deduced that the dinosaur could swim! In a 1980 paper published in *Science*, Massachusetts paleontologist Walter Coombs presented an explanation for an odd sequence of tracks preserved at Dinosaur State Park in Connecticut. This odd sequence was marked by a set of tracks that not only exhibited clear claw marks with very little toe impression, but also ended abruptly. While the sequence had stumped paleontologists for years, Coombs suggested that the individual making these tracks was swimming in shallow water, kick-

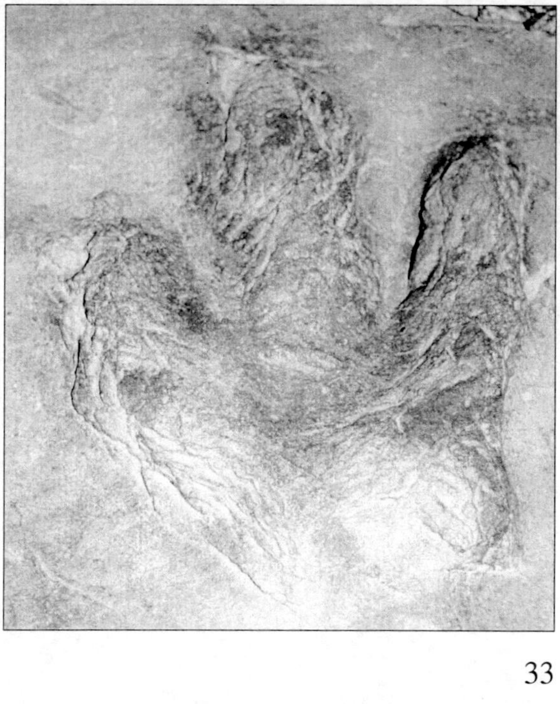

ing the bottom of a lake or sea with the tips of its clawed toes. And, to explain the abrupt ending of the sequence, Coombs suggested that the animal may have been buoyed up while swimming, so that it temporarily lost contact with the lake floor.

As mentioned above, these tracks were discovered in Dinosaur State Park, a Rocky Hill, Connecticut site that remains one of the best-known dinosaur footprint locales in the world. The history of Dinosaur State Park can be traced back to August 24, 1966, when workmen excavating a site that was supposed to become a new State Highway Department testing laboratory hit rock about four meters (12 feet) below the surface. This was expected, but what was preserved in the rock soon made headlines—some 1,500 to 2,000 well-preserved dinosaur footprints. Soon after the discovery, amateur collectors flocked to the site with chisels, hoping to obtain a unique souvenir. Because of this onslaught of visitors, there was an order to halt the construction project. Fences were installed, the tracksite was covered by straw, and troopers were put on guard.

The State of Connecticut quickly rushed to secure a meeting between paleontologists and geologists and the state Finance Commissioner. One of these paleontologists, Yale's John Ostrom, recommended preserving the site. A local newspaper, the *Hartford Courant*, backed the plan. On September 13, less than one month after the site was discovered, Governor John Dempsey designated the site as a state park. Only weeks later Grant Meyer, a young Peabody Museum research associate, supervised the scientific excavation and mapping of the trackway. During this excavation 9,000 square feet of fossil-bearing rock was cleared, treated with plastic sealant, and covered by layers of vinyl and sawdust. Another small trackway was exposed in 1967. Dinosaur State Park officially opened to the public in 1968, and the trackway was designated as a National Natural Landmark by the National Park Service. An air-inflated dome was then erected over 500 of the tracks. Nearly a decade later, in 1977, construction began on a permanent interpretive center to enclose the tracks, which still stands today. This center remains as one of Connecticut's most popular tourist attractions, and also includes a small museum and a variety of paleontology murals and sculptures.

One of the sculptures is of *Dilophosaurus*, an aforementioned possibility for the *Eubrontes* trackmaker. The Dinosaur State Park sculpture was the first *Dilophosaurus* model ever exhibited anywhere in the world, being completed and donated to the park in 1981. Joining the *Dilophosaurus* sculpture is a *Dilophosaurus* mural. It is safe to say that the park promotes this Jurassic theropod as the probable *Eubrontes* trackmaker!

While Dinosaur State Park is a great place to view *Eubrontes* fossils, adventurous visitors to Connecticut can search for their own specimens. *Eubrontes* tracks can be found in the Shuttle Meadow Formation, the East Berlin Formation, and the Portland Formation. The footprints of Dinosaur State Park are found in the East Berlin Formation, as are tracks found near Westfield, Connecticut. However, the richest source for dinosaur footprints in Connecticut is the Portland Formation. This formation yields the famous "brownstone" that is popular both with architects (as a building stone) and fossil hunters. Thousands of dinosaur footprints from this formation have been found near Wethersfield Cove, Middletown, and Portland. Although not directly related to *Eubrontes*, the first recorded dinosaur bones in North America, belonging to the prosauropod *Anchisaurus*, were discovered in the Portland Formation in 1818.

Eubrontes was officially named the state fossil of Connecticut by the State Legislature in 1991. This designation is fitting, since dinosaur footprints are among the best-preserved fossils in Connecticut, and certainly the most important and famous. Luckily, for visitors, these tracks are not only relatively easy to find in the field, but are also easy to locate in museums. In addition to Dino-

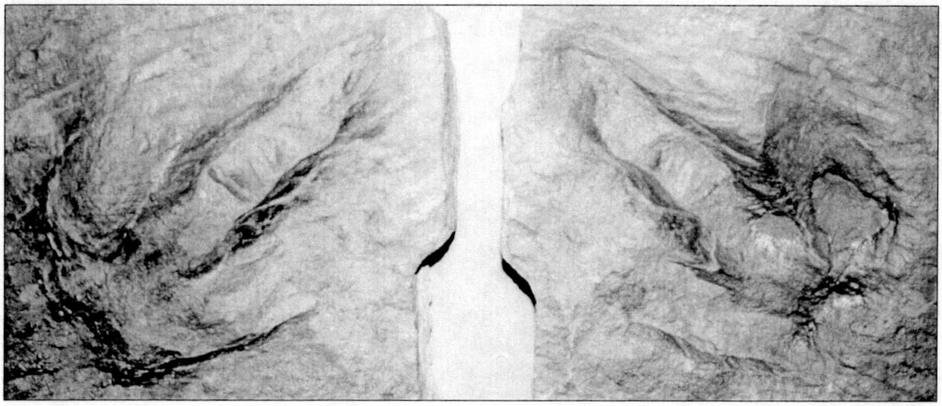

saur State Park, *Eubrontes* tracks are on display at the Pratt Museum in Amherst, Massachusetts, and at the Boston Museum of Science.

Despite its small size, Connecticut is known for its history. To paleontologists and fossil hounds, the Nutmeg State is also known for its prehistory. Foremost among this prehistory are dinosaur footprints, with the most common being the ichnogenus *Eubrontes*. Although the maker of this track remains unknown, it is safe to say that people from all walks of life have enjoyed this track since it was first named by Edward Hitchcock in the 1800s.

FOR FURTHER INFORMATION:

Alter, L., Geology of Connecticut, http://www.yale.edu/ynhti/curriculum/units/1995/5/95.05.01.x.html

Bervoets, F., 2000. New England's Trackways of Connecticut River Valley. Paleozoica, vol. 1, p. 59-60. (viewed at http://www.dinodata.net)

Brusatte, S., 2000. Kids' Corner-*Eubrontes*. Fossil News: Journal of Avocational Paleontology, vol. 6, no. 11, p. 11.

—, 2001a. A Journey Through Deep Time: The Paleontology of New England and Quebec. Fossil News: Journal of Avocational Paleontology, vol. 7, no. 4, p 4-9.

—, 2001b. Fossil Footprints and Paleontology Art Mix: A Trip to Dinosaur State Park. Prehistoric Times, no. 47, p. 30-31.

Little, R.D., 1986. Dinosaurs, Dunes, and Drifting Continents. Alley Geology Publications, Massachusetts.

Olsen, P.E., Smith, J.B., and McDonald, N.C., 1998. Type Material of the Type Species of the Classic Theropod Footprint Genera *Eubrontes, Anchisauripus*, and *Grallator* (Early Jurassic, Hartford and Deerfield Basins, Connecticut and Massachusetts, USA). Journal of Vertebrate Paleontology, 18(3), p. 586-601.

Weishampel, D.B., and Young, L., 1996. Dinosaurs of the East Coast. Johns Hopkins University Press, Baltimore, 275 p.

Dinosaur State Park, http://www.dinosaurstatepark.org

DELAWARE

BELEMNITELLA AMERICANA

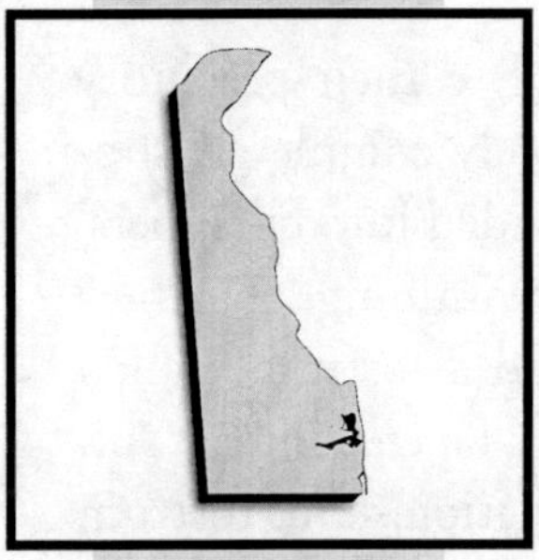

Sparked by the completion of the Erie Canal, the 1820s saw a frenzy of canal-building east of the Mississippi River. In less than ten years, every major lake and river along the East Coast was in some way connected by a canal. One of these waterways was the Chesapeake and Delaware Canal. Completed in 1829, this waterway connected the Chesapeake Bay with the Delaware Bay, enabling eastern products easy transport to western markets. However, little did the builders know that this canal would also spark an interest in fossils.

When the canal was originally built, the construction crews cut through several Cretaceous age marine formations. Although many fossils were destroyed, lots of shells and shark teeth were tossed aside onto giant spoil piles. Today, these spoil piles are prime collecting spots for a particular type of cephalopod, *Belemnitella americana*. Because of its abundance, taxonomic and biogeographic importance, and regional limitation, this belemnite has been designated the official state fossil of Delaware.

Belemnites can trace their ancestry back about 300 million years, to the Mississippian Period. Superficially resembling squid, these molluscs likely evolved from straight-shelled cephalopods. As the shells of these cephalopods became smaller and more robust, the animal's body became lighter and faster. By the dawn of the Cretaceous, belemnites had nearly perfected their hunting and swimming methods, and were abundant across the world.

The seas around present-day Delaware were no exception. With the aging Appalachian Mountains towering over the land to the west, several belemnites, including *Belemnitella americana*, and related cephalopods inhabited the continental shelf regions along the Atlantic.

Belemnites were amazing animals, perfectly adapted to their predatory way of life. Each individual belemnite possessed a soft body completely supported by a cone-shaped internal shell. This shell was divided into chambers, which were filled with gas to maintain buoyancy while swimming. Furthermore, the chambered shell had a blade-like extension and a rostrum, which enclosed the shell and tapered into a conical apex at its termination. This rostrum served as a counter-weight to the buoyancy provided by the chambers, and also was used for protection. The rostrum, also called a guard, is commonly fossilized, while the blade-like extension is rarely found preserved.

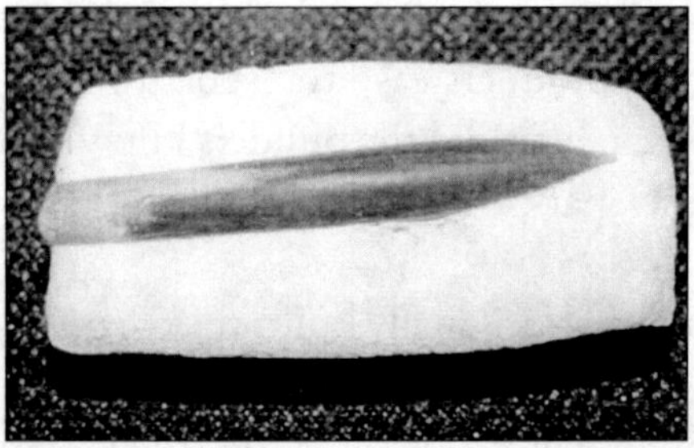

While the shell and rostrum provided protection, several other features of the belemnite were used solely for hunting. Contrary to their simple appearance, belemnites were efficient carnivores that caught and ate small fish, arthropods, and other marine animals. As with any other carnivores, the belemnite evolved several adaptations that aided in the catching and eating of this prey. Among these features were ten hook-laden arms, used for catching prey, and a set of beak-like jaws, used to eat the prey. In addition, belemnites and their cephalopod relatives had a tongue-like radula and advanced brains, which gave them a clear advantage. However, none of these adaptations matched one unusual feature: jet propulsion. Each belemnite possessed a long siphon that shot water out forward, providing the belemnite with backward jet propulsion. Coupled with the tentacles and advanced brains, this feature made the belemnite one of the most efficient and dangerous predators of the Mesozoic. However, nothing gold can stay.

About 65 million years ago, at the sunset of the Mesozoic, all belemnites disappeared. Did they die out with the dinosaurs, possibly from the same cause? This question is debated by many scientists. While some argue that the dinosaurs and belemnites were wiped out because of the same event, likely an asteroid or comet, many think that belemnites went extinct because of competition. When the belemnites died out, bony fishes were moving into upper marine waters. Many of these fish were efficient and fast swimmers, and fed on the same animals as belemnites. In addition, belemnites were also likely the prey of sea birds, which were also gaining in numbers near the dawn of the Cretaceous. Although it cannot be proven conclusively, the rise of faster fish and sea birds may have doomed the belemnites and their cousins, the ammonites.

However, belemnites can still be found in the form of fossils in Delaware and several adjoining states. In Delaware, *Belemnitella americana* is abundantly found in the exposures of the Mount Laurel Formation east of the city of St. Georges in New Castle County, along the banks of the canal. The fine-grained silts and sands of this formation were deposited about 70 million years ago, during the Late Cretaceous, when the belemnites were beginning to disappear.

The most commonly found fossils are fragmented guards, similar in size and shape to a pencil at one end, but flaring at the other. If the guard is well-preserved, it is easy to see the hollow center where the chambered shell was located. However, well-preserved specimens are rare, and often only rod-like broken sections of the rostrum are found. Some specimens are known to reach eight inches, but most are substantially smaller.

There are several places in Delaware where belemnites can be legally collected by private citizens, although commercial collecting is prohibited by the Army Corps of Engineers. The best place to look for fossils is in the dredge spoil piles on the north side of the canal, just east of St. Georges, and east of the north side of the Reedy Point Bridge. The best time to look is after a heavy rain, when pieces of belemnite rostrum may be exposed in the soil.

Joining private collectors at these sites are scientists, who study *Belemnitella americana* to learn more about biogeography. *Belemnitella* is similar to fossils found in parts of New Jersey and England, which shows geologists that either these lands were once connected or that there was a marine faunal exchange between the two areas. However, *Belemnitella americana* is restricted to the Delaware area, a regional limitation that led Delaware Governor Thomas R. Carper to designate this species the official state fossil in July of 1996.

Good *Belemnitella* specimens, along with other fossil belemnites, can be found at several museums along the East Coast, with some of the best specimens housed at the visitors' center at Poricy Park, Monmouth County, New Jersey. However, the best way to see *Belemnitella* fossils is to take a trek to the Chesapeake and Delaware Canal, and examine the spoil piles that hold both American and fossil history.

FOR FURTHER INFORMATION:

Brusatte, S., 2001. Kids' Corner-*Belemnitella*. Fossil News: Journal of Avocational Paleontology, v. 7, no. 10, p. 10.

Doyle, P., 1986. Fossils explained: belemnites. Geology Today, 2: 87-88.

—, 1989. Report on the conservation of the family-group name Belemnitidae. Bulletin of Zoological Nomenclature, 46: 269-270.

House, M. R., 1985. Mollusca (Cephalopoda). in: Atlas of Invertebrate Macrofossils. Murray, J. W., ed. New York: John Wiley & Sons, 148–152.

Lauginiger, E.M., 1978. Cretaceous Fossils from the Chesapeake and Delaware Canal: A Guide for Students and Amateurs. Delaware Geological Survey, 57p.

Pickett, T.E., 1975. Selected Fossil Collecting Locations in Delaware and Minerals in Delaware. Delaware Geological Survey, 2p.

Richards, H. G., 1958. The Cretaceous Fossils of New Jersey. New Jersey Geological Survey Bulletin 61 (2), 139–161.

Windish, D.C. and T.E. Pickett, 1980. Delaware: Its Rocks, Minerals, and Fossils. Delaware Geological Survey, 18p.

Delaware Geological Survey, Delaware State Fossil, http://www.udel.edu/dgs/stfossil.html

Directions: Take Rte. 95 to 202/141 south. Follow 202/141 south of New Castle and take Rte. 9 south. In Delaware City, just after the bridge that crosses the small waterway, turn right onto the small access road. If you cross the C&D canal, you've missed your turn. Take this road past the state park, to the north bank of the canal. Turn left at the end of the road and go under the bridge. Take the left after the bridge up a small hill. You can park anywhere, so find a good sand pile and park there.

GEORGIA

SHARK TOOTH

While Georgia is a charming southern state bordering the ocean today, during much of the past 65 million years, parts of the Peach State were actually underwater. The seas that covered Georgia were a haven for life, which included everything from oysters to fish to mammals. However, one of the most commonly found marine fossils in Georgia is the shark's tooth. Because of its ubiquity and popularity with collectors, the Georgia State Legislature named the shark's tooth the official state fossil.

All sharks are classified as chondrichthyans, which literally means "cartilage fish." This name refers to the internal skeleton of a shark, which is made of cartilage and a superficial layer of hydroxyapatite, the same mineral as makes up true bone. However, unlike bone, this superficial layer consists of thousands of tiny prisms, attached to one another by collagen fibers. This microscopic composition does not favor fossilization, though, and for that reason fossil shark skeletons are rare. Luckily, abundant fossil shark teeth fill this void.

Shark teeth are the most commonly collected vertebrate fossils, and much of this collecting is done in Georgia. There are two reasons for this: the composition of shark teeth and the growth of these teeth. Shark teeth are made of dentine, one of the hardest organic materials known. Dentine also preserves well, making shark teeth common fossils. Secondly, shark teeth are produced on a conveyer belt of gum tissue that lines the surface of the shark's jaws. New teeth are formed within this gum tissue, and are carried closer to the edge of the mouth as the teeth in front of them wear out, at which point they are discarded. Many modern sharks have 3-5 layers of teeth in their jaws at one time, and some species replace their teeth every 7-8 days! This continual formation leads to exorbitant numbers of teeth, all of which can fossilize if the conditions are right. Some estimates show that certain modern sharks can go through 10,000 teeth in their lifetime. While a shark can only produce one fossil skeleton, which is quite rare anyway, it can produce thousands of teeth.

As a group, sharks first evolved during the Late Ordovician, but the earliest remains of these animals are rare and isolated, and include fragmented scales and skin fragments. The first known shark teeth come from the Devonian Period, but it is only during the Mississippian and Pennsylvanian that sharks and their relatives become widespread. The first shark remains known from Georgia are much younger, and date from about 65 million years ago. However, this large missing chunk of time is more than made up for by the abundance and diversity of teeth found in Georgia, which range from small Cretaceous sharks to incredibly large, fierce Miocene species.

Teeth of goblin, mackerel, tiger, and sand sharks are commonly found along Georgia's coast, as are the teeth of what may have been the most fearsome shark to ever live: *C. megalodon.* (The *C.* is an abbreviation for the genus, considered to be *Carcharodon* by some researchers, and *Carcharocles* by others; see below.) *C. megalodon* is an extinct shark that may have reached sizes of 40-100 feet. This size is at least two or three times as long as the modern Great White Shark, which may be a relative of *C. megalodon.* Although no fossilized skeletons of *C. megalodon* have been found, this species is widely known from its teeth, which reached the size of a man's hand!

With teeth this size, it is obvious that *C. megalodon* was a fierce carnivore. Many scientists believe that its diet consisted mostly of whales, but this large shark likely ate whatever it wanted. Estimates show that *C. megalodon*'s jaws could open about six feet (1.8 meters) wide and seven feet (2.1 meters) high, enabling it to swallow large prey. These jaws were loosely attached by ligaments and muscles to the skull, and were lined with the aforementioned teeth, which grew to sizes of 6.5 inches (17 centimeters) and occasionally a bit more. These teeth can be found in Australia, North America, South America, Asia, and Europe.

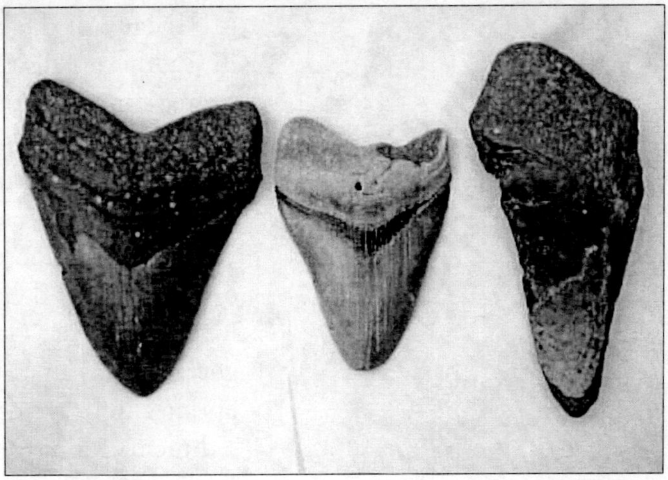

While not related to the state of Georgia, teeth of *C. megalodon* played a large role in the birth of paleontology as a science. In the mid-17th century, Nicholas Steno, a physician to the Duke of Florence in Italy, noticed similarities between the teeth of a modern Great White Shark and "tonguestones," popular curiosities that for centuries had been dug out of cliffs on the island of Malta. In 1667 Steno published a small book, *The Head of a Shark Dissected*, in which he argued that tonguestones were the teeth of long-dead sharks. Today, we know these tonguestones as the teeth of *C. megalodon.*

Despite its relative abundance and popularity with fossil collectors, there is argument over the true classification of *C. megalodon*. The species was first named by Louis Agassiz in 1843, who christened it *Carcharodon megalodon*. However, more modern evidence and recent classifications have shown *C. megalodon* to be directly related to the Great White Shark. This has led a group of scientists to place the species within the genus *Carcharocles*, the same genus as the Great White *(Carcharocles carcharias)*. The debate continues, but many now do regard *C. megalodon* to be a member of *Carcharocles*.

C. megalodon and a plethora of other fossil shark teeth are found in the state's Coastal Plain area, a section of land extending over much of the southern portion of Georgia. Good shark tooth fossils are found in sediments that range from Cretaceous to modern, and are typically found south of Augusta, Macon, and Columbus. The size of teeth found in Georgia varies, from nearly microscopic to the incredibly large sizes of *C. megalodon.*

Many of the better shark teeth found in Georgia are discovered in a small series of formations, including the Ocala Limestone, Clayton Formation, Nanafalia Formation, and the Lisbon Formation. Many of the best collecting areas can be found in the thick Paleocene and Eocene sediments which crop out in Richmond County, Eocene sediments found near Fort Gains and Americus, Oligocene rocks in Laurens County and near Spring and Beaverdam Creeks, and Miocene sediments along the Savannah River near Porter's Landing and Forest Falls Sink. Tooth-bearing sediments are commonly exposed during road and building construction, so serious collectors should frequent work sites to have a better chance of finding rare or well-preserved specimens. For those who are not collectors, Georgian shark teeth can be found in many museums around the world, including the Georgia Southern Museum and the University of Georgia Museum of Natural History.

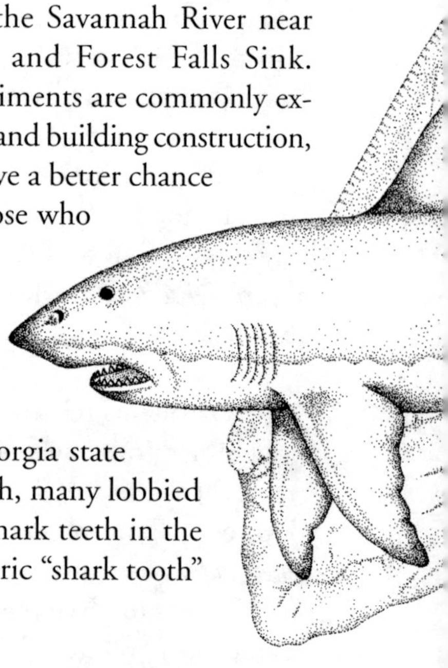

The shark's tooth was officially named the Georgia state fossil in 1976. Before the official naming of the tooth, many lobbied for the adoption of a giant oyster, found alongside shark teeth in the Coastal Plain. However, this plan died, and the generic "shark tooth" was later adopted.

Although the official state fossil is rather nonspecific in name, the importance, diversity, and abundance of shark teeth found in Georgia cannot be denied. Teeth found in the Peach State range from black and grey to brown, reddish-brown, white, and blue, and comprise numerous species, from small, microscopic sharks to the giant *C. megalodon.* Luckily, only a few species of sharks can be found off the coast of Georgia today, and *C. megalodon* is not one of them!

FOR FURTHER INFORMATION:

Applegate, S. P. and L. Espinosa-Arrubarrene, 1996. The Fossil History of *Carcharodon* and its Possible Ancestor, *Cretolamna*: A Study in Tooth Identification. in: Klimley, A. Peter & David G. Ainley (Eds.): Great White Sharks. The Biology of *Carcharodon carcharias*. San Diego, pp. 19-36.

Azuma, Y., 1989. Fossil Tooth of the Lamnoid Shark, *Carcharodon megalodon* from the Kita-Oki Bank, Japan Sea Floor. Bulletin of the Fukui Prefectural Museum, No.3, p.1-7, March 1989.

Cappetta, H., C. Duffin and J. Zidek, 1993. Chondrichthyes. in: The Fossil Record 2. Edited by M.J. Benton, S. 593-609.

Case, G. R., 1967. Fossil Shark and Fish Remains of North America.

—, 1981. Late Eocene Selachians from South-Central Georgia. In Palaeontographica Abt. A, 176, S. 52-79.

Maisey, J. G., 1996. Discovering Fossil Fishes. Henry Holt and Company, New York.

Renz, M., 2002. Meg, We Hardly Knew Ye. Fossil News: Journal of Avocational Paleontology, v. 8, no. 9, pp. 4-7.

Georgia Southwestern State University, Fossil Inventory of Southwest Georgia, http://scimath.gsw.edu/fossils/fossils/shark.htm

National Museum of Natural History, Shark Teeth, http://www.nmnh.si.edu/paleo/sharkteeth/index.html

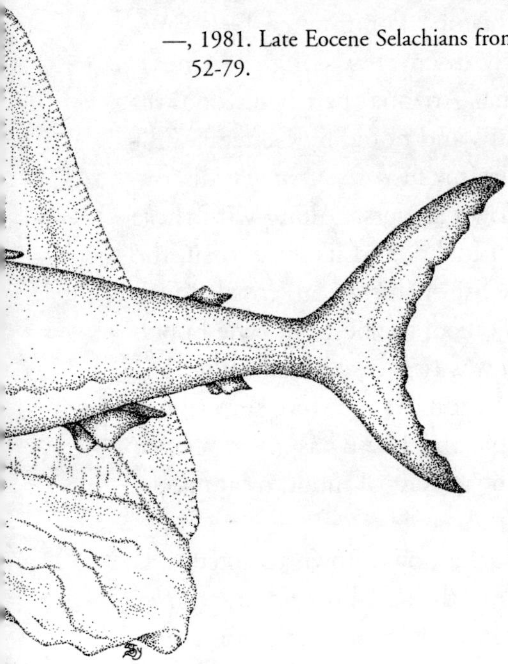

IDAHO

EQUUS SIMPLICIDENS

As you are reading through this book, it is fairly easy to see that several of the state fossils are deeply intertwined with national history. For instance, Delaware's *Belemnitella* is frequently discovered along the Chesapeake and Delaware Canal, Arizona's petrified wood was used by Native Americans and noted by General William T. Sherman, and Virginia's *Chesapecten* was discovered by the first Jamestown settlers. Along with these three states, you can add Idaho, and its state fossil, the Hagerman Horse, to the list of "historical" fossils. This horse, well known from one of the best Pliocene Epoch fossil sites in the world, was first discovered by settlers travelling the famous Oregon Trail. More important, however, are the clues that this horse has given science about the Ice Age and the advent of modern fauna.

The Hagerman Horse, now known to science as *Equus simplicidens*, is one of the best-known early horses ever discovered. It lived in what is now Idaho about 3.5 million years ago, immediately before the last great Ice Age began. At this time Idaho was covered with grassy plains and dotted with ponds and forest stands that received nearly double the precipitation that the area receives today. Mastodons, saber-tooth cats, beavers, otters, fish, frogs, snakes, and waterfowl lived in this lush environment, which was periodically interrupted by flooding caused by the overflow of rainwater in the plethora of rivers, including the young Snake, and prehistoric Lake Idaho. These frequent floods often created sinkholes, which over time became predator traps. It is possible that the massive bonebed of Hagerman Horse fossils discovered in southeastern Idaho represents those of animals once caught in one of these traps.

Regardless of how these fossils formed, the fact remains that paleontologists have learned a lot of new information about pre-Ice Age North America from the cache of Hagerman Horse fossils. The Hagerman is the earliest-known representative of what would eventually become the modern horse. This animal was about the same size as a modern-day zebra or Arabian horse, and if around today would have been quite capable of being mounted. Although a wide variety of sizes are known from fossils, the average Hagerman Horse was somewhere around 110-145 centimeters (43-57 inches) tall at the shoulder and weighed between 110-385 kilograms (385-847 pounds). While it is difficult to deduce from the fossil record, it is likely that the Hagerman Horse possessed a stiff mane and a fur arrangement much like that of modern horses.

It is well-understood that the Hagerman Horse is related to the modern horse. However, scientists are still unclear on exactly what this fossil represents. During the past 70 years this animal has been referred to as a horse, a zebra, and a zebra-like horse. Much of this confusion stems from the skull of the Hagerman Horse. While most of the post-cranial anatomy resembles a true modern horse, the chewing surfaces on the teeth and other details of the skull indicate that the animal was more closely related to the living Grevy's zebra of Africa than to modern true horses. This question is far from

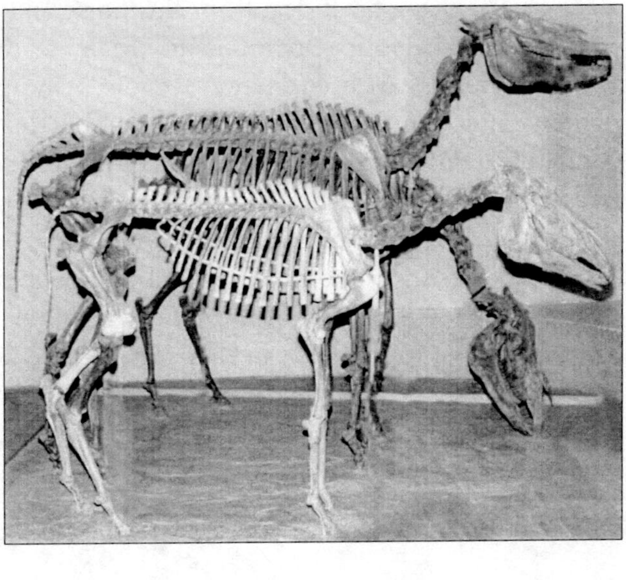

settled, but after initial taxonomic debates, scientists have settled on the scientific name *Equus simplicidens*, a species also known from Texas, Florida, and Nebraska.

Also unclear is how and why this species went extinct. There are two leading hypotheses: disease and climate change. It has been suggested by many that disease may have wiped out the Hagerman Horse, a notion that is easy to envision but hard to prove. Another, better-supported hypothesis is that the environment in southern Idaho, and the entirety of the Great Basin region, was changing from a lush, savanna-like land to a desert-like climate. This is well supported by fossil evidence, and perhaps the Hagerman Horse could not adapt quickly enough to this new environment. Regardless, horses as a group continued to evolve in North America until about 10,000 years ago, at which point they suddenly vanished, perhaps due to human predation and disease. They were later reintroduced in the 1500s by Spanish conquistadors.

Although we are not certain about the Hagerman Horse's evolutionary relationships and extinction, science has learned a great deal from this horse and the animals that shared its environment. Many of these important fossils have been discovered at one of the most famous and significant paleontological sites in the United States: Hagerman Fossil Beds National Monument. This site, located along the Snake River near Hagerman in southeastern Idaho, can trace its roots back to the 19th century, when pioneers travelling the Oregon Trail noted strange bones near the cliffs that overlooked the river. However, it would not be until 1928 that the site became known to science. In that year, a cattle rancher named Elmer Cook discovered a cache of strange bones 450 feet above the river. Not recognizing the bones as belonging to any type of cattle, Cook showed samples to Dr. H.T. Stearns of the U.S. Geological Survey, who later passed them on to Dr. James W. Gidley of the Smithsonian Institution.

Gidley identified the bones as belonging to an extinct horse, and began to excavate the area near where the bones were found in 1929. The first year's excavations provided three tons of fossils, which were shipped back to the Smithsonian. The successful yield led the Smithsonian to mount several more excavations in the 1930s. By the end of the decade, some 120 horse skulls and 20 complete skeletons were excavated from an area dubbed "Horse Quarry," one of several bone sites excavated. Since that time, the Smithsonian has exchanged some of these skeletons with other museums across the world, and additional scientific expeditions have been conducted. These expeditions and donated fossils have resulted in more than 200 published scientific papers on the Hagerman Horse. Today, the Hagerman site is known as one of the best Pliocene Epoch fossil locales in the world.

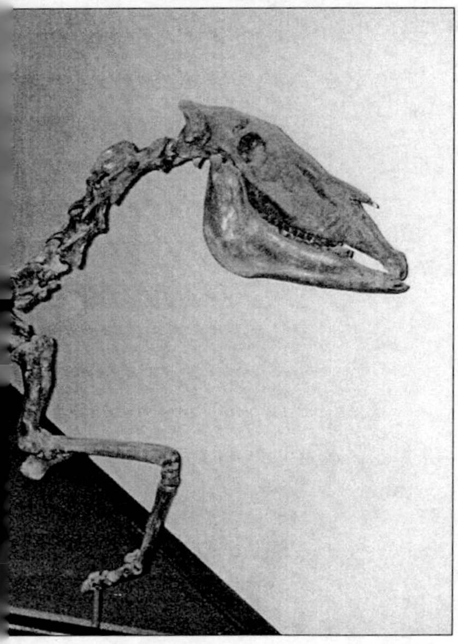

After finding the site, scientists were at a loss trying to explain exactly why so many skeletons of horses and other animals were preserved there. One early explanation was that the quarry area was once a watering hole where the bones of injured, old, and ill Hagerman Horses accumulated after death. However, it is now thought that an entire herd of horses likely drowned in an attempt to ford a flooded river and were swept away by the current. Their bodies were quickly buried beneath the soft sand on the river floor, and over the next several thousand years, turned to stone. Then, some 14,500 years ago, the Hagerman Valley was formed during the well-documented Bonneville Flood, which swept through the Snake River valley when Lake Bonneville (the predecessor of Great Salt Lake) breached its banks and drained to the north. This flood exposed countless layers of the Glenns Ferry and Tuana Formations, rock units which contain the preserved remains of the Hagerman Horse.

However, thousands of years after this flood, erosion still endangers the Hagerman Horse remains. As late as 1987 a landslide, caused by an irrigation project, sent tons of fossil-bearing rock cascading into the Snake River. This landslide angered Hagerman residents, who joined with other Idaho citizens to promote the Hagerman Horse as a state fossil and called for preservation of the site. Only a short time later, on March 16, 1988, the horse was named the state fossil, and the Governor proclaimed May 28, 1988 "Hagerman Horse State Fossil Day." On November 18 of that same year, several of the fossil sites became part of Hagerman Fossil Beds National Monument. This site, one of the most popular tourist destinations in Idaho, is one of only three sites in the National Park System to preserve parts of the historic Oregon Trail.

Because of the discoveries of Elmer Cook and James Gidley, the efforts of the Hagerman citizens, and the foresight of the National Park Service, the Hagerman Fossil Beds are preserved from erosion and illegal collecting, safe for generations to come. Fossil remains of this famous horse can be seen not only at the Hagerman site, but also at countless museums around the world, including the Carnegie Museum in Pittsburgh, the Field Museum in Chicago, and the Texas Memorial Museum in Austin. While the Pliocene Epoch is preserved at several fine fossil sites around the United States, it is at Hagerman that scientists have gotten the best look into the strange pre-Ice Age world of North America.

FOR FURTHER INFORMATION:

Brusatte, S., 2000. Kids' Corner-Hagerman Horse. Fossil News: Journal of Avocational Paleontology, vol. 6, no. 7, p. 11.

Idaho State University, Hagerman Horse Quarry, http://imnh.isu.edu/digitalatlas/ida_geology/idaho_parks/hagerman/horse_quarry/horse_quarry.html

National Park Service, Hagerman Fossil Beds, http://www.aqd.nps.gov/grd/parks/hafo/

Only a few miles from Malad Gorge, on U.S. Highway 30 south of Interstate 84, is the town of Hagerman, where the park headquarters for Hagerman Fossil Beds National Monument, 221 N. State Street, (208) 837-4793, is located. A visitors center offers information and fossil displays. The 4,281 acre monument is composed of 600-foot-high bluffs west of Hagerman across the Snake River.

ILLINOIS

TULLIMONSTRUM GREGARIUM

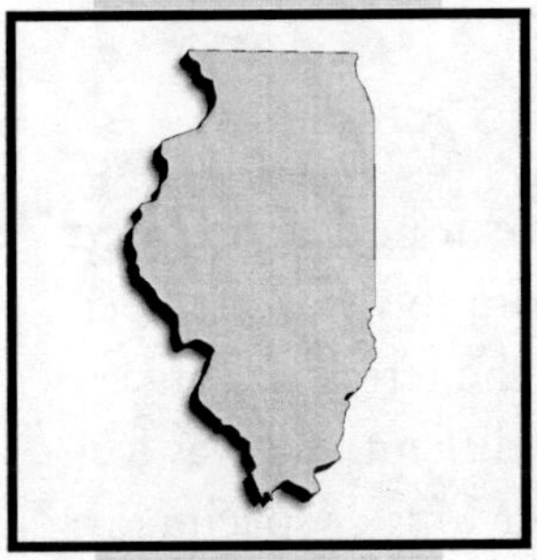

Because its fertile soils preserve no known dinosaurs, paleontologists and fossil collectors often overlook Illinois. However, in all reality, the Land of Lincoln boasts one of the most diverse, abundant, and interesting fossil records in the United States. Despite the fact that Illinois has largely been resurfaced by glaciers and the majority of its rock units have been scoured, a host of fossils are found in the state, including wonderful Ordovician, Silurian, Devonian, Mississippian, Pennsylvanian, Cretaceous, Tertiary, and Quaternary specimens. Among these fossils, the most famous is undoubtedly the enigmatic Tully Monster, a strange worm-like animal known from the renowned Mazon Creek fauna of northern Illinois.

The Tully Monster is like nothing living today, and can trace its ancestry back to the Pennsylvanian Period, about 300 million years ago. During this time, most of the ancient continents had converged to form a supercontinent known as Pangaea. When the existing continents collided to form Pangaea during the Middle Pennsylvanian, parts of North America were compressed and uplifted, forming tall mountains. Mountains to the east of what was to be Illinois created conditions of atmospheric circulation that produced extensive rainfall. This rainfall, in turn, created vast seas, swamps, and deltas. One of the largest of these delta and swamp systems was located in Illinois, which was positioned only 5° from the equator.

These swamps were chock full of dense vegetation, and provided a perfect habitat for countless species of plants and animals. Along the margins of these swamps was located a vast tropical sea, which was also a haven for marine life. Occasionally, plants and animals from the coastal swamps would be washed into the sea. A unique mode of anaerobic preservation fossilized many of these organisms. Today, fossils of a plethora of plants and animals can be found in central Illinois near what is today called Mazon Creek. One of the most interesting is the Tully Monster.

The Tully Monster was a worm-like soft-bodied animal. It often preserves as outlines and flattened forms in nodules of ironstone found at isolated Mazonian fossil sites. Unlike some of the fossils known from the Mazon Creek area, the Tully Monster lived in the tropical sea, and like the terrestrial organisms discussed above, it became trapped under layers of sediment and was fossilized. This fossilization has preserved a plethora of Tully Monster specimens. However, this abundance has not helped scientists in figuring out exactly what the Tully Monster was.

Often described as worm-like, the Tully Monster is an elongated, bilaterally symmetric fossil animal ranging in length from about 7.5 centimeters (three inches) to just over 30 centimeters (one foot). The body of the Tully is divided into three distinct regions: the anterior head end, the trunk, and a posterior end with a bluntly pointed tail and lateral fins. The Tully Monster may have been segmented, but recent work suggests that it was not.

Perhaps most amazing about the Tully Monster is its long proboscis, located on the anterior (head) end of the animal. At the end of this trunk-like organ was a primitive "jaw" that contained up to 14 small, sharp teeth. There is no evidence that the proboscis was continuous with the throat, but instead it probably represents a muscular organ used to pass food to the mouth. Even this interpretation is controversial. Near the middle of the body was a transverse bar characterized by swellings on both ends. These circular swellings may have been the Tully's sensory organs. However, a recent study conducted by B.S. Beall suggested that the transverse bar may have been a copulatory organ or a benthic support structure. Like the proboscis, the interpretations of this transverse bar are also ambiguous. Finally, the posterior (tail) end was divided into three triangular fins. These fins were likely swimming adaptations. Also, it is thought by most that the Tully Monster was an active carnivore. Its fins and predatory habits are about all scientists agree on, though.

It is obvious from the above description that the Tully Monster is a strange, enigmatic animal. After its original naming in 1966, scientists have argued over the phylogenetic relationships and ancestry of this worm-like creature. Originally, in 1969, Eugene S. Richardson, the man who named the Tully Monster, concluded that it was related to no known phylum. However, in 1979, Merrill Foster argued that the Tully Monster was related to heteropod gastropods, writing that similarities in eye form, tooth shape, and overall morphology link the two. Later cladistic analyses by Beall and F.R. Schram only convoluted the picture further, suggesting that the Tully Monster was either an annelid or closely related to conodonts or molluscs. As of today, the relationships of this animal are poorly understood at best, and even better specimens may not resolve the debate over its proper phylogenetic position.

The Tully Monster was first found by Francis Tully in 1958. Tully, an avid fossil collector from Lockport, Illinois, discovered the strange fossil near Braidwood. He took the specimen to the Field Museum of Natural History in Chicago, where none of the staff scientists could identify it. Because of its odd appearance, the animal was dubbed "Tully's Monster" by museum scientists, and the name eventually stuck. In 1966, Eugene Richardson, then Curator of Fossil Invertebrates at the Field Museum, gave the animal a proper scientific name in a landmark paper published in *Science*. Richardson dubbed the enigma *Tullimonstrum gregarium*, which literally means "Tully's common monster."

Although they are very odd, Tully Monsters are not rare. Today, they are commonly found in Will and Kankakee Counties, at Chowder Flats near Morris in Grundy County, and at the Sunspot Mine near Astoria in Fulton County. Like other Mazon Creek specimens, Tully Monsters became fossilized when the animal was buried quickly by fine clays. Molecule-by-molecule, the living tissue was mineralized and eventually became preserved in iron siderite concretions. These concretions, which are often dark red or brown, are commonly found throughout central Illinois, in an area enclosed by the towns of Essex, Morris, Wilmington, and Ottawa. Because of this unusual mode of preservation, soft-bodied animals, organisms that are rarely fossilized under normal conditions, are often found in the Mazon Creek area. All-in-all, well over 500 species of plants and animals have been discovered in this famous biota.

The Tully Monster is restricted to the Essex Fauna of Mazon Creek, one of two distinct faunal facies known from the area. The Essex Fauna preserves a brackish to marine environment which was dominated by jellyfish, fish, molluscs, crustaceans, worms, and the Tully Monster. Little stratigraphic information about this habitat is known because many of the fossils were exposed as a byproduct of strip coal mining, which destroyed the original rock layers. However, unlike the stratigraphy, the paleoecology of this area is well-known and well-studied.

The Tully Monster officially became the state fossil in 1989, after Illinois State Geological Survey paleontologist Donald Mikulic lobbied the State Legislature. After Mikulic was called a "communist" by a legislator opposed to the designation, the Tully bill passed the General Assembly and was signed into law. Today, good specimens of the Tully Monster can be seen at the Field Museum, the Illinois State Museum in Springfield, the Burpee Museum of Natural History in Rockford, and the Cernan Space Center at Triton College.

While Illinois can boast no dinosaurs, the Land of Lincoln is home to an amazingly diverse fossil record. Foremost among these fossils is the enigmatic Tully Monster. Although its phylogenetic relationships are unknown, the Tully Monster is prized by fossil collectors and scientists alike, who marvel at its amazing preservation and strange suite of features.

FOR FURTHER INFORMATION:

Beall, B.S., 1991. The Tully Monster and a New Approach to Analyzing Problematica, pp. 271-285, in Alberto M. Simonetta and Simon Conway Morris (eds.), The Early Evolution of Metazoa and Significance of the Problematic Taxa, Proceedings of the International Symposium, Cambridge University Press, Cambridge.

Brusatte, S., 2000. Kids' Corner-Mazon Creek. Fossil News: Journal of Avocational Paleontology, vol. 6, no. 4, p. 11.

—, 2002. Kids' Corner-*Tullimonstrum gregarium.* Fossil News: Journal of Avocational Paleontology, vol. 8, no. 8, p. 10.

Foster, M.W., 1979. A Reappraisal of *Tullimonstrum gregarium.* in M.H. Nitecki (ed.), Mazon Creek Fossils, Academic Press, New York, pp. 269-302.

Johnson, R.G., and Richardson, E.S., 1969. Pennsylvanian Invertebrates of the Mazon Creek Area, Illinois: the Morphology and Affinities of *Tullimonstrum gregarium.* Fieldiana Geology, 12(8): 119-149.

Konecny, J., 2000. The Mazon Creek Nodules. Fossil News: Journal of Avocational Paleontology, vol. 6, no. 9, p. 4-7.

Leary, R.L., 1987. The Tully Monster: Mystery Animal of Illinois. The Living Museum, 49: 26-27.

Mikulic, D.G., and Kluessendorf, 1998. Illinois' State Fossil-*Tullimonstrum gregarium.* Illinois State Geological Survey Geobit 5. (view online at http://www.isgs.uiuc.edu/servs/pubs/geobits-pub/geobit5/geobit5.html)

Richardson, E.S., 1966. Wormlike Fossil from the Pennsylvanian of Illinois. Science, 151: 75-76.

Shabica, C.W., and Hay, A.A., 1997. Richardson's Guide to the Fossil Fauna of Mazon Creek. Northeastern Illinois University Press, 308 p.

Wysocki, G., 1998. Mazon Creek Madness! Fossil News: Journal of Amateur Paleontology, vol. 4, no. 4, p. 4-5.

Illinois State Museum, Illinois State Fossil,
http://www.museum.state.il.us/exhibits/symbols/fossil.html

KENTUCKY

BRACHIOPOD

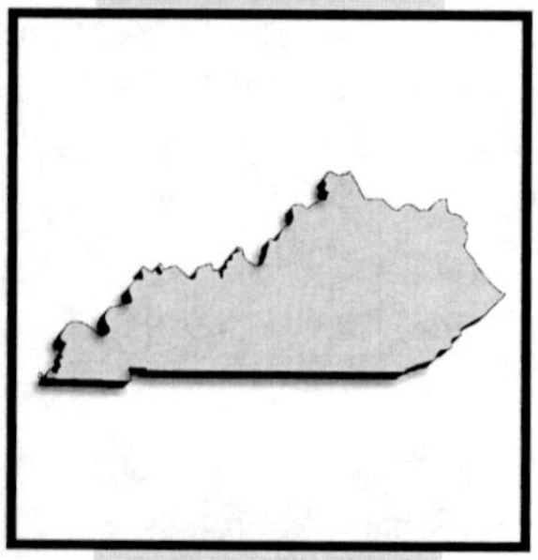

The home of bluegrass, horse breeding, and rolling hills today, Kentucky was covered by water during most of the Paleozoic. As would be expected, these seas were teeming with invertebrate life, including corals, bryozoans, trilobites, and crinoids. All of these invertebrates are common fossils in Kentucky. However, none comes close to the ubiquity of brachiopods, a phylum of bivalved marine invertebrates that were once the most common animals in Paleozoic seas. Because of their diversity and the fact that they are commonly found in Kentucky sediments, the brachiopod has been named the official state fossil of Kentucky.

Like other dual-valved invertebrates, brachiopods possessed two distinct shells. However, the similarity ends right there. Unlike clams and other members of the Class Bivalvia (which are in the Phylum Mollusca), brachiopods (Phylum Brachiopoda) have shells that are dissimilar, with one shell often larger than the other. Brachiopod shells are symmetrical from side to side—the plane of symmetry passes through the valve, not between them as in a clam—so that the top and bottom shells look different. The internal anatomy of brachiopods, both extinct and modern, is also very different from that of clams and oysters. Brachiopods have a coiled feeding organ called a lophophore, which is occasionally preserved in some Devonian brachiopods found in Kentucky. This organ occupies about two-thirds of the shell interior, and functions essentially as a pump, bringing in oxygenated water and food particles and generating currents that remove wastes. Most brachiopods also attach to the sea floor with a fleshy, muscular stalk known as the pedicle. The body cavity is hollow, containing a rudimentary digestive system and some kidney-like organs, but lacking separate circulatory and respiratory systems.

There are two major classes of brachiopods: inarticulate and articulate. Inarticulate brachiopods were likely the first to evolve, and can be found in rocks dating to the Cambrian Period. Inarticulate brachiopods open and close their valves with muscles and do not rely on the socket-and-teeth hinge mechanism that is a characteristic of articulate species. Inarticulate Devonian species in Kentucky are also found attached to other organisms, including other brachiopod shells and corals. Furthermore, the bottom valve of inarticulate brachiopod species is often cemented to a firm surface, with the upper dorsal valve of the shell open for feeding.

Articulate brachiopods, on the other hand, have two shells that are different sizes. The larger shell is called the pedicle valve, and contains a hole through which the fleshy pedicle emerges. Furthermore, this pedicle valve is lined with projections called teeth, which fit into sockets on the opposite shell, known as the brachial valve. Articulate brachiopods come in a variety of shapes, and their shells often show external ornamentation. Some are lined with costae, or fine ridges, on the outer surface of the shell. Some articulate brachiopods may have even been brightly colored, to stand out in their environment.

Brachiopods were (and are) benthic feeders, living the duration of their lives on the bottom of the sea. Furthermore, several brachiopod species also were attached to the ocean floor by spines, which helped stabilize the animals in muddy sediment. In addition to the aforementioned lophophore, brachiopods also use short, hair-like cilia to feed. Generally, brachiopods open their shells just enough to allow a stream of water to enter, while the lophophore and cilia trap and capture microscopic food particles.

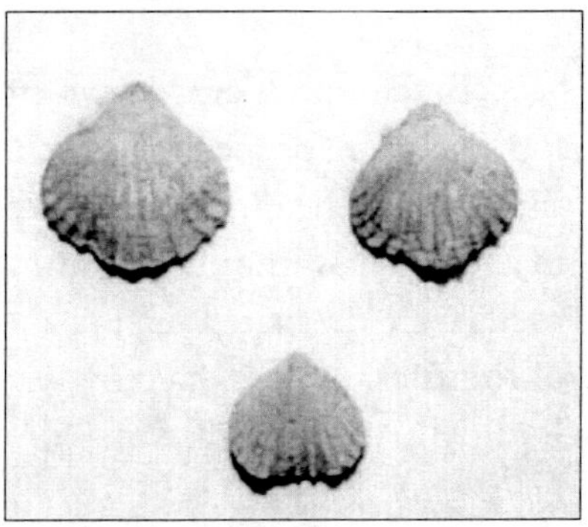

Brachiopods first evolved during the Cambrian, but only became widespread in the Ordovician. They are well known from Ordovician, Silurian, Devonian, and Mississippian sediments in Kentucky. However, most brachiopod species died out about 245 million years ago, at the end of the Permian Period. Although they still exist, brachiopods have never regained their Paleozoic diversity. Exactly what caused this decline? Many brachiopods make their homes in very cold water, either in polar regions or at great depths in the ocean. Perhaps a cooling episode at the end of the Permian disrupted their preferred environment. Or, possibly, an asteroid or comet collided with the earth, leading to the widespread Permian extinction in which 90% of marine invertebrates died out. The answer may never be known. However, nearly 300 living species of brachiopods are present today, mostly in cold environments.

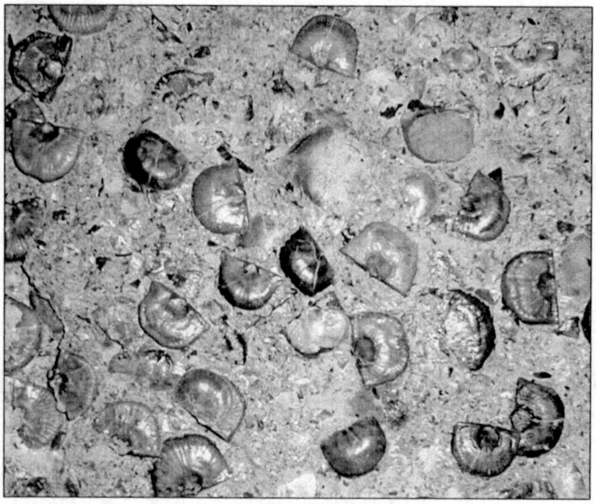

While 300 species may seem like a large number, thousands of extinct, fossil brachiopod species have been named. The plethora of brachiopod fossils can be attributed to the construction of the shell, which is composed of calcium carbonate or phosphate, both of which are easily fossilized. In Kentucky, brachiopods are the most common fossils.

Brachiopods are known from much of the Paleozoic in Kentucky, including a plethora of specimens from the Ordovician, Silurian, Devonian, and Mississippian. The oldest rocks exposed at the surface in Kentucky are Ordovician in age, and commonly contain well-preserved brachiopod fossils. Most of these rocks were deposited during the first half of the Ordovician, when the entirety of Kentucky was covered by a shallow tropical sea. Some common brachiopods found in the Ordovician sediments include *Rafinesquina, Platystrophia*, and *Hebertella*.

Silurian rocks, although not as widespread as Ordovician limestones, are exposed at the surface in the Knobs region of Kentucky. As during the Ordovician, Kentucky was covered by shallow seas throughout much of the Silurian, which means that numerous Silurian fossils are found in the Bluegrass State. Later, during the Devonian, much of Kentucky was still covered by shallow seas, although some very low lands may have been emergent at times in the central portion of the state. However, these dryland areas found themselves submerged once again during the Late Devonian, when the entire state was covered with deep, poorly-oxygenated water. Dark, organic-rich muds were deposited, producing the famous Devonian black shales that contain not only fossil brachiopods, but also oil. Finally, during the Mississippian, most of Kentucky found itself under shallow seas once again. Rocks deposited in these seas commonly outcrop in the Mississippian Plateau of the state, and contain several good brachiopod species.

Although it is obvious that the brachiopod fossil record is diverse in Kentucky, the most famous and popular specimens come from a small area along the Ohio River. Known as the Falls of the Ohio, this area encompasses several acres on both the Indiana and Kentucky sides of the river. On the Indiana side, a beautiful state park has been developed, which includes a 16,000-square-foot interpretive center containing many brachiopod fossils.

The Falls of the Ohio were deposited during five different intervals in the Devonian period. The uppermost layer, the youngest of the five, is called the *Paraspirifer acuminatus* Zone, named for the fossil brachiopods commonly found in the unit. The second layer is known as the Fenestrate Bryozoan-Brachiopod Zone, which pays homage to the many bryozoans and brachiopods found in the layer. Finally, the third layer is called the *Brevispirifer gregarius* Zone, and contains many brachiopod fossils. Six distinct orders of articulate brachiopods and two orders of inarticulate brachiopods are found at the Falls of the Ohio. Collecting at the Falls of the Ohio is prohibited, since it is a protected area. But good brachiopod fossils can be legally collected at many sites in Kentucky, including road cuts where the Fort Payne and Borden Formations are exposed.

Because of the ubiquity of these brachiopods, teachers and 8th grade students at Myers Middle School in Louisville—excited after reading an account of Montana's official state fossil—campaigned for the designation of a brachiopod as Kentucky's state fossil. However, with so many brachiopods known from so many different rock units, which one would be chosen? The Kentucky State Legislature took the middle road when they designated the generic "brachiopod" the official state fossil on July 15, 1986 with the passage of Act 488, Section 1.

While Kentucky boasts no dinosaurs, the Bluegrass State can brag about the diversity of invertebrate fossils found within its borders. Among these invertebrates, the brachiopods constitute the most common phylum of fossils found. It is more than fitting, then, that the generic "brachiopod" is known as the official state fossil.

FOR FURTHER INFORMATION:

Brusatte, S., 2001. Kids' Corner-brachiopod. Fossil News: Journal of Avocational Paleontology, v. 7, no. 12, p. 20.

Grabowski, G.J., Mississippian System, Contributions to the Geology of Kentucky, http://pubs.usgs.gov/prof/p1151h/miss.html

Greb, S.F., 1989. Guide to the Progression of Life. Falls of the Ohio State Park, 44 p.

Kindle, E.M., 1901. The Devonian Fossils and Stratigraphy of Indiana. Indiana Dept. of Natural Resources.

McCormick, T., 1993. Fossils Galore. Outdoor Indiana Magazine, (can be viewed online at http://www.fallsoftheohio.org/history.html)

Moore, R.C., ed., 1965. Volume H, Brachiopoda. in: Treatise on Invertebrate Paleontology, University of Kansas and Geological Society of America, 927 p.

Nettelroth, H., 1889. Kentucky Fossil Shells. Kentucky Geological Survey.

Falls of the Ohio State Park, http://www.fallsoftheohio.org/brochures/brachiopod.html

University of California, Berkeley, Introduction to the Brachiopoda, http://www.ucmp.berkeley.edu/brachiopoda/brachiopoda.html

University of Kentucky, Fossils of Kentucky, http://www.uky.edu/KGS/coal/webfossl/fosslbig.htm

LOUISIANA

PETRIFIED PALM WOOD

Although several states list petrified wood as their official stone, Louisiana is one of two states that list this common and well-known fossil as their official state fossil. While dinosaur bones are loved by children and marine fossils are known throughout much of the country, it may be petrified wood that is the most often-collected fossil in the United States. This collecting is a result of the beauty of petrified wood, which is often preserved in a dazzling array of colors. It is no surprise, then, that petrified wood is a common jewelry stone, and is a frequently-seen object in souvenir stores in Louisiana.

What we call petrified wood is nothing more than fossilized tree stems, trunks, and branches, and is often composed of a variety of minerals, including silica, sulfates, sulfides, oxides, silicates, carbonates, and phosphates. These minerals permeate the wood after it dies and is buried, and over time turn the woody material to stone. Often, this petrified wood is found within two types of strata. First, it is found within accumulations of volcanic ash and breccia. Two famous examples are the *in situ* petrified tree trunks found in and near Yellowstone National Park in Wyoming and Florissant Fossil Beds National Monument in central Colorado. Secondly, petrified wood is also commonly found in sandstone, siltstone, and shale. These sedimentary rocks are the remnants of sand, silt, and mud deposited by rivers and streams, which have since hardened under the pressure of deep burial. The famous Petrified Forest National Park (see the chapter on Arizona) and the Oligocene strata of Louisiana are examples of this second mode of preservation.

The petrified wood found in Louisiana formed in the same way as most other petrified wood specimens. After the tree died and was buried, dissolved silica within the groundwater attached itself to the lignin and cellulose of the wood. Over time, a layer of monomeric silicic acid formed on the exposed woody tissues. This acid dehydrated into silica gel, while additional layers of acid attached to this gel, eventually filling the wood with the silica gel. This process caused the wood to rapidly lose water, which converted the silica gel into amorphous silica. Within the next several million years, this silica, now known as opal, further dehydrated and crystallized into microcrystalline quartz, or chert. During this process, various minerals and other materials sometimes became incorporated into the silica gel, causing it to change color or appearance.

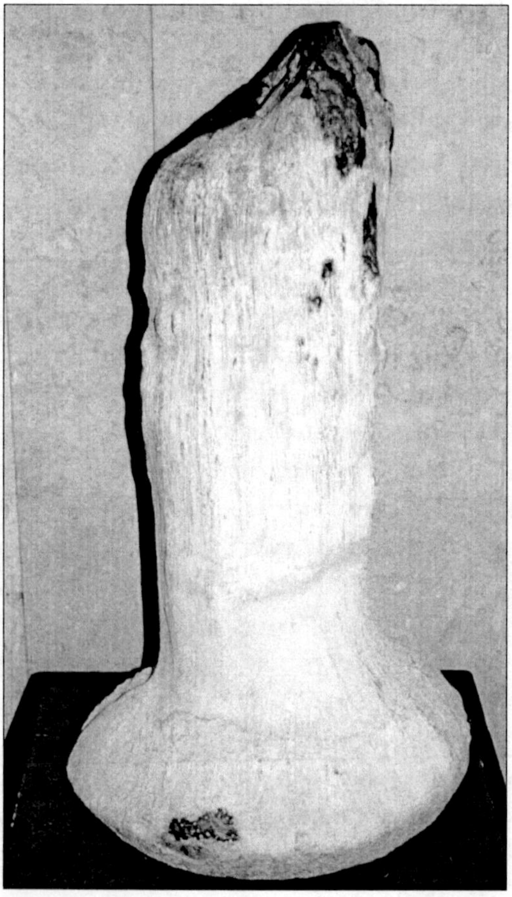

Several types of petrified wood are known from the region, including three distinct varieties from the Tertiary (mostly Oligocene) strata that crop out in Louisiana and parts of Texas. The first type is rather nondescript, and consists of silicified wood that possesses a very recognizable woody structure. It is known simply as "silicified wood." The second type is known as "massive silicified wood." In these specimens, the destruction of the structure of the silicified wood has resulted in a massive opal or chert that lacks any visible indication of its original origin. Finally, the third type of petrified wood known from Louisiana is commonly called "petrified palm wood." This type of wood is likely the best known, and has been designated the official state fossil.

Petrified palm wood is basically a group of fossil woods that contain prominent rod-like structures within the regular grain of the silicified wood. Often, these rod-like structures show up as spots, tapering rods, or continuous lines. Although they have long since been altered, these rod-like structures are remnants of sclerenchyma bundles that once comprised part of the woody tissues that gave the wood its vertical strength. This wood is a favorite of geologists, rock collectors, and tourists because of its structure and colors. Because of these colors, it is often used as a jewelry stone, and because it is composed of silica it is hard enough to polish and withstand the wear of normal use.

The "petrified palm wood" found in Louisiana represents three known species of monocotyledonous trees—*Palmoxylon microxylon, Palmoxylon cellulosum*, and *Palmoxylon lacunosum*. (As with Arizona's *Araucarioxylon, Palmoxylon* is a paleobotanical form genus, indicating that the wood resembles that of modern palms, but may or may not be originally from trees closely related to them.) Three more *Palmoxylon* species are known from neighboring Mississippi, and another from Texas. These trees grew in what is now Louisiana some 30 million years ago, during the Oligocene Epoch. During the Oligocene, the Gulf of Mexico's shoreline was much farther north, which explains why specimens from these trees are discovered in the more northern Rapides, Natchitoches, and Sabine Parishes. Together, these three parishes and part of Texas comprise the Toledo Bend area, one of the best areas to find petrified wood in the world. While it is hard to determine much about the individual trees based on petrified wood alone, we can compare the specimens to modern palm trees, which are still common in the Gulf of Mexico area. Modern palms are distinguished from most other trees in that they have no branches and their leaves are all clustered at the top. The same is assumed for Oligocene palms.

Petrified palm wood has long been a popular item with collectors in Louisiana. Although it is not known who first discovered this wood, scientists have discovered stone tools made of petrified wood at Paleoindian sites, proving that

Native Americans knew about and used the wood. The first European-American known to collect petrified wood in Louisiana was L.C. Johnson, who found several specimens in Rapides Parish in 1885. Geologist Frank Knowlton used Johnson's specimens and other material he collected to write the first published scientific descriptions of palm wood from Louisiana. Since that time the only other published study of Louisiana palm wood was written by Edward Berry in 1916.

Petrified palm wood was officially designated the Louisiana state fossil by Act No. 362, which was passed as Senate Bill No. 155 by the Louisiana Legislature in 1976. The governor of Louisiana approved this act on July 31, 1976. While it is still unknown why petrified palm wood was named the state fossil, Louisiana geologist Paul Heinrich wrote about its designation. Citing a different 1988 article, Heinrich's article stated, "it is interesting to note that in 1976, while the state legislature was considering the naming of the state fossil, one of the legislators nominated Senator Edgar Mouton of Lafayette for that honor. After the Senator declined in deference to 'age, rather than beauty,' the (Miocene) petrified wood was declared the winner by acclamation."

At the time the fossil was officially designated, the Catahoula Formation, the rock unit in which petrified palm wood is frequently found, was considered to be of Miocene age. Currently, it is thought to be older, and has been referred to the Oligocene. However, this dating is also still somewhat controversial.

Despite this controversy, though, petrified palm wood remains one of the most popular fossils from the American south, and is one of the most frequently-collected fossils in the United States. Its popularity has manifested itself in both Native American stone tools and modern jewelry. While some may think that petrified palm wood is not the most exciting or dramatic of fossils, its importance and popularity make it an excellent choice for that state honor.

FOR FURTHER INFORMATION:

Berry, E.W., 1916. The Flora of the Catahoula Sandstone. U.S. Geological Survey Professional Paper Number 98, p. 227-251.

Brusatte, S., 2002. Kids' Corner-petrified palm wood. Fossil News: Journal of Avocational Paleontology, v. 8, no. 1, p. 9.

Heinrich, P.V., Petrified Palm Wood, http://www.intersurf.com/~heinrich/Palm.html

Knowlton, F.H., 1888. Description of Two Species of "Palmoxylon" from Louisiana. Proceedings of the United States National Museum, p. 89-91.

McMackin, C.E., 1984. Petrified Wood from East to West; Some We've Liked Best. Lapidary Journal, vol. 37, no. 11, p. 1582-1588.

Paine, W.W., and Meyerhoff, A.A., 1968. Catahoula Formation of Western Louisiana and Thin Section Criteria for Fluviatile Depositional Environment. Journal of Sedimentary Petrology, vol. 38, no. 1, p. 92-113.

Pope, D.E., 1988. History of the Louisiana Geological Survey. in Socolow, A.A., ed., The State Geological Surveys: A History, American Association of State Geologists, 499 p.

Zeitner, J.C., 1988. Louisiana's Agatized Palm. Lapidary Journal, vol. 42, no. 4, p. 43-46.

MAINE

PERTICA QUADRIFARIA

The state motto of Maine translates as "I lead." Therefore, it is more than fitting that the state fossil of the Pine Tree State is one of the earliest known land plants, a species that literally led the first terrestrial plant lineage to a new home on terra firma. Living during the Devonian Period, *Pertica quadrifaria* was as much an alien on land as the first astronauts were on the moon. However, because of the boldness of this plant and its relatives, land plants are extremely successful and diverse today.

Along with the Tully Monster of Illinois, *Pertica quadrifaria* is likely one of the most localized and unique of the state fossils. Although a similar species has been found in Scandinavia, *Pertica* is only known from a very small region of Maine, a state that otherwise has very few fossils. This dearth of fossils is due to the geological makeup of Maine, which is rich in metamorphosed rocks and deeply interwoven with granite. Metamorphism is, unfortunately, very damaging to fossils. However, for one shining moment during the Devonian, about 390 million years ago, a chance eruption of a volcano preserved intact a vast colony of early terrestrial plants, helping paleontologists reconstruct not only the early landscape of Maine, but the evolution of land plants. This plant, *Pertica quadrifaria*, may be among the most important of all state fossils.

During the Devonian, Maine would have looked absolutely nothing like it does today. Instead of a mountain-laden, snow-filled land, Maine would have been much warmer, and home to a teeming ecosystem of early land plants and animals. However, land plants as a group did not evolve either in Maine or during the Devonian. Instead, their lineage can be traced back to the Silurian, the shortest, but perhaps one of the most significant, of all geological periods.

According to the estimates of most scientists, the earliest terrestrial plants appeared on land sometime during the Silurian Period, likely around 425 million years ago. For millions of years before this, though, plants were abundant in the seas. Why, then, didn't they colonize the land earlier? The answer lies in the biological makeup of plants. While there is no gravity to speak of in water, plants must grow against gravity on land. Therefore, any successful land plant must evolve a hard structural or hydrostatic pressure support system to keep it upright. Secondly, plants cannot afford too much evaporation from moist surfaces, so they must evolve some type of waterproofing. Third, land plants need some sort of anchor to keep them in the ground. Finally, terrestrial plants also must be able to transport water and other minerals throughout their systems, as they do not have the luxury of constant submersion in a watery medium. Because of these needs and many more, it took plants millions of years to evolve the necessary structures to colonize the lands, and even the earliest land plants were much different than are most plants around today.

It is unknown exactly how terrestrial plants evolved, although paleontologist John Raven has concocted a believable scenario. According to his hypothesis, some green algae may have evolved a water transport system between adjoining cells to move food quickly around the plant. While living in seasonal environments, some algal colonies may have temporarily grown up into the air, which would have increased the level of photosynthesis, as light intensity in the air is higher than that in the water. As these plants grew upwards into the atmosphere, the water transport system became more specialized, as did the uppermost tips of the plant, which were modified into structures used solely for photosynthesis, much like pseudo-leaves. Once these systems became modified completely, the plant may have been able to venture onto land. This hypothesis is nicely crafted and believable, but nearly impossible to prove.

Silurian land plants were relatively simple. However, the Devonian is characterized by a very strong radiation of more advanced land plants. Four major groups of these plants can be distinguished: the rhyniophytes, the zosterophyllophytes, the lycopods, and the trimerophytes. *Pertica* was a trimerophyte.

As a rule, trimerophytes have a complex water transport system, technically known as a water vascular system, and terminally positioned sporangia standing in dense clusters. These sporangia were modified structures used for releasing spores, simple reproductive structures still seen in modern ferns. Many trimerophytes could reach a considerable height. The best estimates show that *Pertica* could grow to sizes of about six feet, with some scientists suggesting that this plant may have even reached nine feet. More important than their size, though, is the fact that trimerophytes likely gave rise to all subsequent groups of land plants except for the lycopods (clubmosses).

Pertica lacked leaves and roots, and most of its body consisted of branching stems that contained photosynthesis-performing cells throughout their length. Vascular tissue was present, forming a solid central bundle in the center of its large stem. Unlike other closely-related plants, *Pertica* and other trimerophytes possessed branches that were unequal, forming a main stem with several smaller lateral branches. These lateral branches often were shortened to form bushy webs of small, closely-spaced branches. *Pertica* also may have possessed enations, small flaps of tissue very similar to leaves, but lacking the water vascular system seen in all modern leaves. *Pertica* also bore sporangia at the tips of its branches.

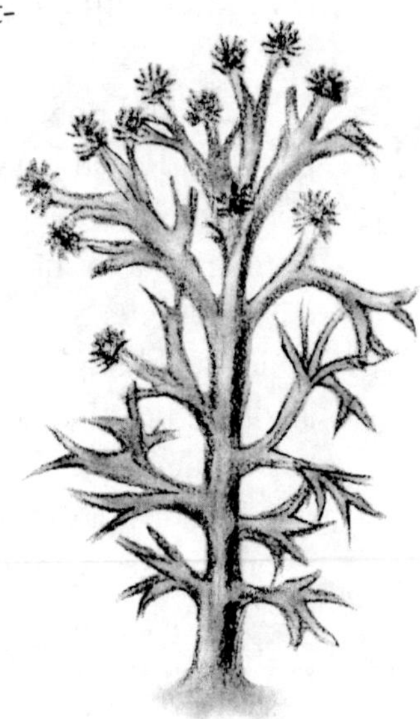

The first fossils of *Pertica* were discovered in 1968 in the rocks of the Trout Valley Formation in Baxter State Park near Mount Katahdin. Four years later, in 1972, it was officially named *Pertica quadrifaria*, meaning "long pole or rod with four forked tips," by paleontologists A. Kasper and H. Andrews.

Today, fossils of *Pertica* are still found abundantly in the Trout Valley Formation in north-central Maine. Why is this fossil so abundant in this part of Maine, but unknown from any other area? The answer lies in both the lifestyle of *Pertica* and its environment. Most scientists believe that *Pertica* lived in freshwater or brackish marshes, and probably grew in large colonies. Judging by the type of rock it is found in, the vast Trout Valley *Pertica* colonies grew near an ancient volcano, which occasionally erupted huge quantities of ash, covering the large *Pertica* colonies and preserving their delicate tissues. Today, *Pertica* specimens are commonly found eroding out of stream banks. Because of the fact that the volcanic eruptions preserved large colonies, *Pertica* fossils are quite abundant when they are found. However, because it is the state fossil, all collectors wishing to collect any samples of *Pertica* must obtain a permit from the the Maine Land Use Regulation Commission.

Because of its obvious paleontological importance and its extremely localized presence, the Maine State Legislature declared *Pertica quadrifaria* the official state fossil in 1985. Today, collectors can either obtain a permit to hunt *Pertica* specimens, or view them in local museums, including the Maine State Museum, and at Bowdoin College in Brunswick. William Forbes, one of the scientists who published on *Pertica* in the 1970s, recently donated much of his scientific collection to Bowdoin, where some of it is now on display.

While it may not be the most ideal state for fossil collecting, Maine may have the most unique and localized state fossil. However, the true importance of *Pertica* lies in the fact that it lies directly on the lineage that leads to modern land plants. Long before Maine adopted "I lead" as its state motto, *Pertica* led its fellow plants onto an alien world.

FOR FURTHER INFORMATION:

Kasper, A., and Andrews, H., 1972. *Pertica*, a new genus of Devonian plants from Northern Maine. American Journal of Botany, v. 59, p. 892-911.

Kasper, A., Andrews, H., and Forbes, W., 1974. New fertile species of *Psilophyton* from the Devonian of Maine. American Journal of Botany, v. 61, p.339-359.

Kasper, A., Gensel, P., Forbes, W., and Andrews, H., 1988. Plant Paleontology in the State of Maine—A Review. Maine Geological Survey, Studies in Maine Geology, v. 1, p. 109-128.

Stewart, W.N. and Rothwell, G.W., 1993. Paleobotany and the Evolution of Plants. Cambridge University Press, Cambridge.

Maine Geology, Maine's State Fossil, http://www.state.me.us/doc/nrimc/pubedinf/factsht/paleo/stfossil.htm

University of California at Berkeley, Introduction to the Trimerophytes, http://www.ucmp.berkeley.edu/plants/trimerophytophyta/trimero.html

University of Munster Research Group, History of Paleozoic Forests: The Earliest Land Plants, http://www.rrz.uni-hamburg.de/biologie/b_online/kerp/seite3.html

MARYLAND

ECPHORA GARDNERAE GARDNERAE

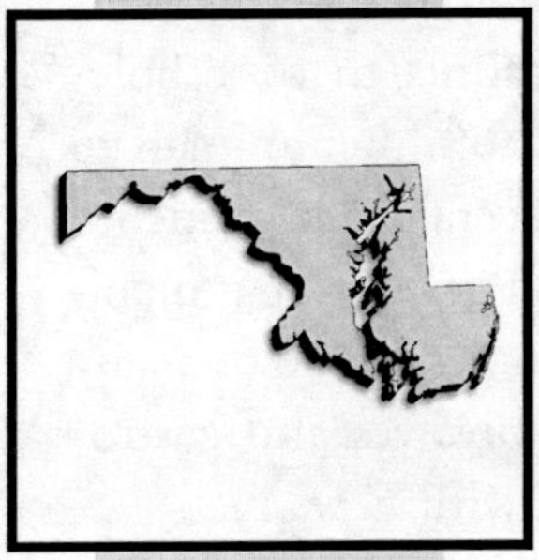

Although it is quite small, Maryland has a diverse fossil record. This record stretches from large Mesozoic dinosaurs to small Miocene snails. It is fitting, then, that Maryland is one of the few states which have designated both an official state fossil and a state dinosaur. The state dinosaur will be discussed later. Here we will focus on the official state fossil, the snail *Ecphora gardnerae gardnerae.*

During the Miocene Epoch, some 5-12 million years ago, much of Maryland was covered by water. Inhabiting the continental shelf were several species of molluscs, including bivalves (such as *Chesapecten jeffersonius*, the state fossil of Virginia) and gastropods. One such gastropod was *Ecphora gardnerae gardnerae.*

Like modern snails, *Ecphora* was characterized by a spiral shell, into which the snail could withdraw when threatened. Although it is not present in the fossil record, like all snails, *Ecphora* also possessed a flattened foot, used for locomotion, and a head with prominent tentacles. Also like modern snails, *Ecphora* would have undergone body torsion during early development. Snails begin life as bilaterally symmetrical organisms. However, during the larval stage the visceral mass is twisted 180°, so that posterior structures such as the anus are brought forward, toward the mouth. There are several advantages to this torsion, including the fact

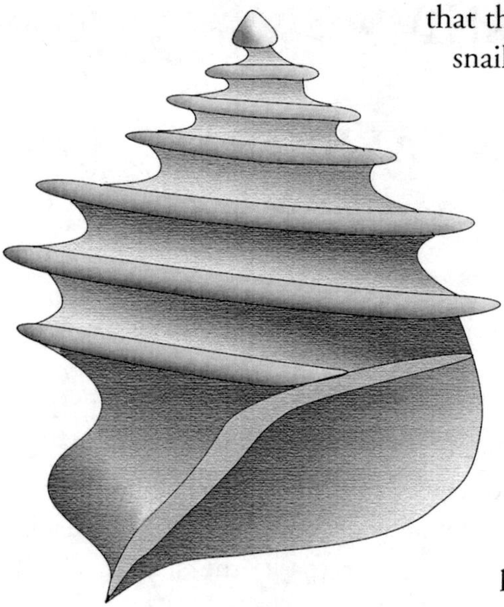

that the shell opening is rotated, enabling the snail's head to withdraw inside.

Most modern snails are herbivorous. However, *Ecphora* was a predator. It probably preyed on other snails, bivalves, crustaceans, marine worms, and corals, and used its tongue-like radula to bore holes into the shells of other animals to feed on the visceral mass. Evidence of this method of hunting is manifested in boreholes, found on many fossil Miocene bivalves and gastropods that coexisted with *Ecphora*.

In addition to its predatory lifestyle, *Ecphora* also differed from most other snails in three distinct characteristics. First, *Ecphora* shells are russet in color, which contrasts with the more common white color of most other fossil molluscs. Secondly, four strong costae, or "protruding ribs," are present on *Ecphora* shells. Third, *Ecphora* has a moderately wide, hollow cone-shaped feature along the axis of coiling, a structure known as an umbilicus.

The genus *Ecphora* probably first appeared during the Oligocene Epoch, around 30 million years ago. However, *Ecphora gardnerae gardnerae* first evolved during the Miocene. Several other *Ecphora* species are represented in Oligocene, Miocene, and Pliocene sediments and seem to change often over time, making them valuable index fossils—species that, when found, help to place a particular stratum in time. However, towards the end of the Pliocene, all *Ecphora* fossils disappear from the fossil record. What was the reason for this extinction? Many scientists point to a prolonged period of extreme global cooling, which may have killed not only *Ecphora*, but many other marine molluscs.

Some two million years after it became extinct, *Ecphora* became one of the first fossils to be illustrated in a scientific publication. A drawing of the snail appeared in a 1770 edition of Martin Lister's *Historiae Conchyliorum* (History of Conchs). In this book, *Ecphora* was labeled "a marilandia," or a shell "from Maryland." The probable collector of the illustrated shell was Hugh Jones, a rector of Christ Church in Port Republic, Maryland, from 1696 to 1701. An avid naturalist and fossil hunter, Jones regularly collected and sent many local plants, animals, and fossils to England.

However, it took several more years until *Ecphora* was officially named in a scientific publication. The author of the paper christened Lister's snail *Ecphora quadricostata*, and based the description on a collection of fossil snails found in Virginia. However, in the ensuing century, paleontologists discovered that there were differences between *Ecphora quadricostata* and the snails commonly discovered in much of Maryland. Because of these differences, Smithsonian Institution paleontologist Druid Wilson proposed the name *Ecphora gardnerae* for the Maryland shells in 1987, honoring Dr. Julia Gardner of the U.S. Geological Survey. Soon after, two other scientists further subdivided the genus *Ecphora* and introduced several subspecies, one being *gardnerae*. This designation gave birth to the name *Ecphora gardnerae gardnerae*, with the second "gardnerae" designating the subspecies.

This confusion over naming also involved the Maryland General Assembly, which officially named *Ecphora quadricostata* the state fossil in 1984. In light of the name change, the General Assembly passed Chapter 688, Section 13-311 on October 1, 1994, officially designating *Ecphora gardnerae gardnerae* the state fossil.

Regardless of the nomenclatural confusion, *Ecphora* is still considered one of the most important and renowned fossils found in Maryland. There are several collecting localities for the various species of *Ecphora* in the state, including the Calvert, the Choptank, and the St. Mary's Formations. However, *Ecphora gardnerae gardnerae* is only found in the St. Mary's Formation, named for St. Mary's County, the site of the first *Ecphora* discovery in the 1600s.

The best public collecting site for *Ecphora* is undoubtedly the famous Calvert Cliffs, north of Little Cove Point. These cliffs are located in the largest fossil-bearing deposit of Miocene marine sediments exposed on the East

Coast of North America. While most of the cliffs are located on private property, small-scale collecting is permitted along a small section of beach at Calvert Cliffs State Park, a recreation area owned by the state of Maryland. Remains of sharks, whales, porpoises, mammoths, birds, and of course, marine invertebrates are common at Calvert Cliffs, although overcollecting has left some areas bare.

The sediments that make up Calvert Cliffs were laid down 10 to 20 million years ago, when the Atlantic Coast was repeatedly submerged beneath the sea. Studies of fossil plants and animals, including *Ecphora*, indicated that the ocean which covered the area was warm and shallow, with cypress swamps lining the shores. The climate was somewhat warmer than today, although the temperature changed drastically at the end of the Pliocene. After millions of years of existence, the ocean retreated and what was once the ocean bottom is now the Calvert Cliffs.

These cliffs extend for more than 30 miles from just north of Chesapeake Beach to Drum Point, rising to more than 100 feet in height. Although all three of the aforementioned formations are exposed along the cliffs, the St. Mary's Formation yields some of the most impressive fossils, including a glut of molluscs, many of which are collected between South Beach and Governor Run. To date, some 400 species of molluscs have been identified from the deposits, with about 44 of them still living.

Because of Calvert Cliffs and other collecting areas, *Ecphora* has become a renowned and popular fossil in Maryland. In fact, the Maryland senator who sponsored the bill to make *Ecphora gardnerae gardnerae* the official state fossil collected fossil snail shells along the Chesapeake Bay as a boy with his brother, who went on to become a geologist. *Ecphora gardnerae gardnerae* is even the symbol of the Paleontological Research Institution of Ithaca, New York. While Maryland has a diverse fossil record, few specimens can compete with the popularity of *Ecphora gardnerae gardnerae*.

FOR FURTHER INFORMATION:

Ashby, W.L. and Parrish, M.A., Fossils of Calvert Cliffs. Calvert Cliffs State Park (informational pamphlet).

Germon, R. N., Ward, L. W., and Ray, C. E., 1988. *Ecphora*—important fossil from the Miocene strata on the Chesapeake Bay. The Maryland Naturalist, 31(1): 25-33.

Petuch, E. J., 1988. New species of *Ecphora* and Ecphorine Thaidids from the Miocene of Chesapeake Bay, Maryland, U.S.A. Bulletin of Paleomalacology, 1(1): 1-16.

Ward, L.W., 1993. Evolution of a name, or what species of *Ecphora* is that again? American Paleontologist, Paleontological Research Institution, Ithaca, NY, 1(2): 1-2.

Ward, L.W. and Blackwelder, B.W., 1975. *Chesapecten*, a new genus of Pectinidae (Mollusca: Bilvalvia) from the Miocene and Pliocene of eastern North America. U.S. Geological Survey Professional Paper 861, 24 p.

Ward, L. W. and Gilinsky, N. L., 1988. *Ecphora* (Gastropoda: Muricidae) from the Chesapeake Group of Maryland and Virginia. Notulae Naturae, Academy of Natural Sciences of Philadelphia, No. 469, 21 p.

Wilson, D., 1987. Species of *Ecphora* in the Pungo River Formation. in C. E. Ray (ed.), Geology and paleontology of the Lee Creek Mine, North Carolina, II: Smithsonian Contributions to Paleobiology, 61:21-29.

Maryland Geological Survery, Maryland's Official State Fossil Shell, http://mgs.dnr.md.gov/esic/fs/fs6.html

MASSACHUSETTS

Dinosaur tracks

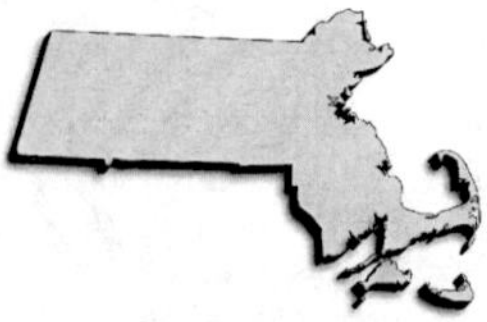

Much of the early history of the United States can be traced to Massachusetts, a small New England state which has seen much of our nation's past, from the dawn of the Pilgrims to the death of the Kennedys. It is no surprise, then, that Massachusetts also plays a major role in the prehistory of the United States. Although its fossil record isn't as vast as that of Wyoming or Montana, or as diverse as that of Illinois or Ohio, Massachusetts is known for one particular type of fossil: dinosaur tracks. These dinosaur tracks are so popular and well-known that the state legislature declared them the official state fossil in 1980.

The dinosaur tracks of Massachusetts can be traced back to the dawn of the Jurassic Period, some 200 million years ago. At this time, the supercontinent of Laurasia was beginning to split into two, forming what would become the Atlantic Ocean. Along the nascent break was a giant rift valley, which was repeatedly downdropped by a series of powerful earthquakes. Swift-flowing rivers from the mountains to the west gradually deposited over 16,000 feet of sediment into the valley, which created large open mudflats that were home to countless species of dinosaurs. Sheets of the mineral mica, left by receding floodwaters, covered these mudflats periodically, preventing the mud from sticking to the feet of the dinosaurs as they walked. This allowed for the preservation of thousands of dinosaur footprints, which over time, turned into stone.

Several types of dinosaur prints are found in Massachusetts, including *Eubrontes, Grallator, Anchisauripus*, and *Otozoum*. Of these, the ichnogenus *Eubrontes* is one of the most common. (Trace fossils are called ichnofossils, and their names ichnotypes, to distinguish them from biological species.) This print likely belonged to a mid-sized, bipedal, predatory dinosaur. A typical *Eubrontes* track ranges from 40-50 centimeters (16-20 inches), and is marked by three long toes, often with evidence of a heel. Based on these measurements, the *Eubrontes* trackmaker must have been over one meter high at the hips and 5-6 meters long. One possible culprit is *Dilophosaurus*, whose body proportions fit this description nicely, and is also known from the same general time period.

Another common print is *Grallator*, an ichnogenus named after the Family Grallae, a group of birds that includes the herons and storks. Although *Grallator* does not belong to a bird, as originally thought, this notion was not far off—given that birds are now thought to be the direct descendants of small, meat-eating dinosaurs. *Grallator* tracks closely resemble those of large birds, measuring an average of 2-6 inches in length and containing three toes, often with impressions of a fourth. They were undoubtedly laid down by a small primitive theropod, possibly *Coelophysis, Podokesaurus*, or a close relative.

Resembling the *Grallator* prints is an enigmatic track designated as *Anchisauripus*, literally meaning the "track of *Anchisaurus*." However, in recent years, several scientists have concluded that these tracks were instead made by theropods, not prosauropods like *Anchisaurus*. *Anchisauripus* tracks measure anywhere from 6-10 inches long, slightly larger than *Grallator*. It has even been suggested that these *Anchisauripus* tracks may actually represent the footprints of large individuals of the *Grallator* stock. Possibly, the *Anchisauripus* tracks represent adults, while *Grallator* prints are those of juveniles. This hypothesis has been suggested, but has yet to be conclusively proven.

A fourth common track in Massachusetts is *Otozoum*, a track originally described as that of an amphibian. Later, it was described as belonging to a prosauropod. However, today many researchers believe it to be crocodylian in origin.

With the glut of dinosaur footprints known from Massachusetts, it is no wonder that the first dinosaur footprints ever recorded were discovered in the Bay State. In the spring of 1802, a young boy named Pliny Moody hit rock while plowing his father's field near South Hadley, Massachusetts. He examined this rock, and noticed that it contained a set of 30-centimeter-long tracks, which Moody described as "three-toed, like a bird's." He installed the slab on his doorstep, and it soon began attracting local visitors, including religious authorities. One authority claimed the prints belonged to "Noah's raven," a bird once supposedly housed in Noah's Ark.

The tracks also drew another local authority, Amherst College professor and president Edward Hitchcock, who viewed the slab at the Moody household. He agreed that the prints were laid down by a bird, but rejected the notion of Noah's raven. Instead, he wrote, a large prehistoric bird had made these tracks thousands, if not millions, of years ago. Hitchcock later purchased the slab, and it is still housed at Amherst College.

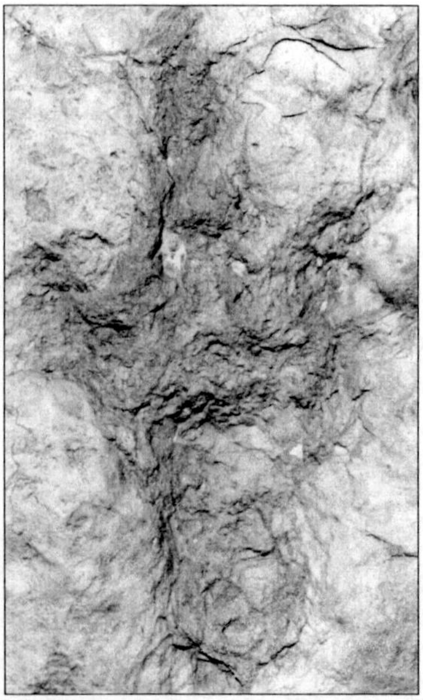

Years later, in 1835, the town of Greenfield, Massachusetts began a street-resurfacing project, using locally-quarried stones from Turners Falls. However, on closer inspection, town officials noticed that some of the stones bore markings similar to turkey tracks. These tracks were brought to the attention of local physician James Deane, who wrote a letter about the discovery to Hitchcock. After viewing casts of these tracks, Hitchcock wrote, "as soon as I saw the specimens, I perceived the phenomena worthy of careful research." Because of this find, Hitchcock began a comprehensive study of the footprints of Massachusetts and Connecticut, taking two years to crisscross the Connecticut River valley looking for good specimens. In 1836 he published his first paper on the tracks, in which he attributed them to extinct giant flightless birds. He named several of them "Ornithichnites," meaning "stony bird tracks." However, he soon began to recognize specimens that looked much different. Over time, Hitchcock distinguished and named several types of tracks, attributing them to birds, mammals, and reptiles. None were attributed to dinosaurs, animals which were still unknown when Hitchcock began his research. It is only today that we realize the true makers of these tracks.

Since the time of Hitchcock, numerous other dinosaur tracksites have been discovered in Massachusetts, including those in South Hadley, Holyoke, Turners Falls, and Springfield. The majority of these prints are in the Jurassic Shuttle Meadow, East Berlin, and Portland Formations.

One of the most important of these recent discoveries has been that of Mt. Tom near Holyoke, Massachusetts. The Mt. Tom site, near a ski resort, was first documented by John Ostrom in 1972. Ostrom described the site as having a track-bearing surface of 40 meters by 20 meters (150 by 60 feet), and populated by at least 134 three-toed footprints ranging in length from nine centimeters to 35 centimeters. Ostrom asserted that these tracks showed clear evidence of herding behavior. He identified 28 separate trackways, and approximately 70% are oriented in nearly parallel courses, suggesting that the dinosaurs who made them were travelling together. This site, known as the Dinosaur Footprint Reservation, can be visited today.

Near Mt. Tom is the city of Holyoke, the site of a strange track discovery. In March of 1996, while digging for a catfish pond, landowner Gary Gaulin hit rock. As a science buff, Gaulin knew to look for footprints. However, what he found still amazed him: hundreds of three-toed dinosaur tracks. As of 2000, Gaulin had excavated some eight feet of the site, exposing countless more tracks.

Near the original Moody Farm is a third popular footprint site, Nash Dinosaurland in South Hadley. The site's original owner, the late Carlton Nash, discovered his first dinosaur track on the property in 1933. When he was older, he bought the land, and transformed the area into one of the world's largest commercial dinosaur footprint quarries. Nash operated a small gift shop, and even sold tracks to dignitaries like General George Patton. However, Nash recently passed away, leaving the future of the site in doubt.

Finally, dinosaur tracks were discovered in Granby, Massachusetts by Bill Gringas in 1968. When digging through rocks in preparation for building a deck, Gringas noticed some strange markings on a slab of shale he had just uncovered. Today, the Gringas home is the site of the Granby Dinosaur Museum. Visitors are welcome in Granby, as they are in several other museums which house Massachusetts tracks, among them the Pratt Museum of Amherst College, the Boston Museum of Science, and the Springfield Science Center in Springfield. Really adventurous visitors can try their hand at digging for their own tracks, although the best prints are often located on private property, so permission is required.

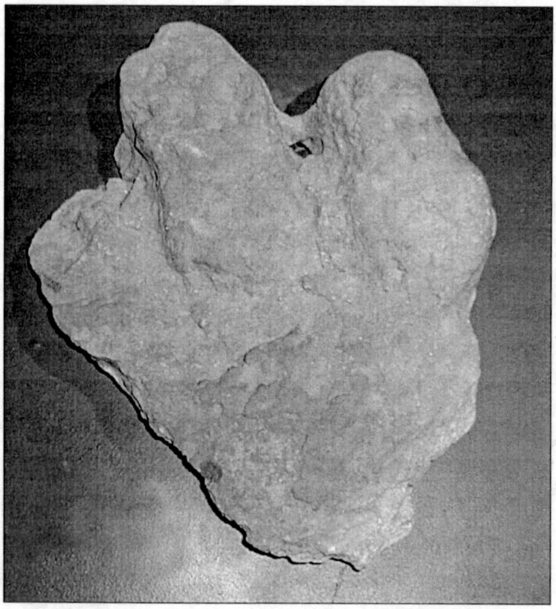

Massachusetts, a small New England state, is a mother lode for both history and prehistory buffs. In one day, visitors can check out the site of the Boston Tea Party, tour the U.S.S. Constitution, visit a dinosaur track museum, and hunt for their own footprints. Such is the joy of this small, charming state, which has officially designated the generic "dinosaur track" as its state fossil.

FOR FURTHER INFORMATION:

Bervoets, F., 2000. New England's Trackways of Connecticut River Valley. Paleozoica, vol. 1, p. 59-60. (viewed at http://www.dinodata.net)

Brusatte, S., 2001. A Journey Through Deep Time: the Paleontology of New England and Quebec. Fossil News: Journal of Avocational Paleontology, vol. 7, no. 4, p 4-9.

Hitchcock, E., 1836. Ornithichnology: Description of the Footmarks of Birds (Ornithichnites) on New Red Sandstone in Massachusetts. American Journal of Science, 29:301-340.

—, 1858. Ichnology of New England, A Report on the Sandstone of the Connecticut Valley, Especially its Fossil Footmarks. W. White Publishers, Boston, 232 p., 60 pls.

Lockley, M.G., 1991. Tracking Dinosaurs. Cambridge University Press, Cambridge.

Olsen, P. E., Smith, J. B., and McDonald, N. G., 1998. Type material of the type species of the classic theropod footprint genera *Eubrontes, Anchisauripus*, and *Grallator* (Early Jurassic, Hartford and Deerfield basins, Connecticut and Massachusetts, USA). Journal of Vertebrate Paleontology, v. 18, no. 3, p. 586-601.

Olsen, P., Triassic-Jurassic Footprint Project, http://www.ldeo.columbia.edu/~polsen/nbcp/tjfp.html

Smith, J.B., Smith, J.R., and Sweeney, J.P, 1996. Mt. Tom Revisited: a New Fossil Horizon at Dinosaur Footprint Reservation, Northampton, Massachusetts. Journal of Vertebrate Paleontology, 16(3):67A.

Weishampel, D.B., and Young, L., 1996. Dinosaurs of the East Coast. the Johns Hopkins University Press, Baltimore, 275 p.

Amherst College, Hitchcock Ichnology Collection, http://www.amherst.edu/~pratt/collections/hitchcock.html

University of Massachusetts, Connecticut River Geology,
http://www.bio.umass.edu/biology/conn.river/firstimpressions.html

MISSISSIPPI

Basilosaurus and Zygorhiza

You will recall from the section on Alabama that Eocene whales are well known from the Gulf Coast area. These early whales, which have helped scientists reconstruct the origins and evolution of the Order Cetacea, are so prevalent and important that they have been designated by two states as official state fossils. While Alabama named *Basilosaurus cetoides* as its official fossil, Mississippi has chosen to designate two of these important whales: *Basilosaurus* and *Zygorhiza kochii*.

While many other groups of mammals, such as the elephants and saber-tooth cats, receive more attention from paleontologists and fossil buffs, the whales make up one of the most intriguing groups to have ever arisen. What makes the whales so interesting is that they evolved from terrestrial hoofed mammals, and only through drastic changes in their morphology have they achieved the structures necessary for life in the sea. Although their derivation is somewhat controversial, it is thought that whales evolved from a terrestrial group of primitive basal artiodactyls. Much of the information on the artiodactyl-whale connection has been deduced from fossils found in Mississippi.

During the Eocene Period, some 45 million years ago, the lands that make up modern-day Mississippi were under water. This aquatic environment was teeming with life, and was a haven for invertebrates, fish, and archaeocetes, or primitive whales. Two of the best-known and most important primitive whales are *Basilosaurus* and *Zygorhiza*. Both of these whales show structures that unambiguously link them to the artiodactyls, and help scientists understand the transition from land to water.

Evolving an aquatic form from a terrestrial body is no easy task, and making this transition required a host of evolutionary changes. Among these changes were a backwards and upwards shift of the nostrils, loss of the rear limbs, a streamlining of the shape, change of the forelimbs into flippers, the addition of flukes for swimming, the loss of most hair, and the modification of senses for use in the water. The first mammals to show these changes were the archaeocetes ("ancient whales"). As a general rule, these archaeocetes possessed tiny heads and pointed snouts with differentiated teeth, much like those of most land mammals. In contrast, modern whales either lack teeth—relying instead on sieve-like baleen plates—or have teeth that are virtually all identical in shape and size. These archaeocetes, *Basilosaurus* and *Zygorhiza* among them, also had nostrils near the tip of the nose and lacked a blow hole.

By the Late Eocene, some 45-35 million years ago, archaeocete whales had spread to many parts of the world, including the Mississippi area. During this time, the Gulf of Mexico extended much further north, and covered the majority of modern Mississippi. Although the notion is hard to prove, it is thought that many archaeocete whales came to the Mississippi area to give birth, an event which may still have taken place on or near land early in whale evolution.

Basilosaurus is the better-known of the two Mississippi state fossils, and resembled a giant sea serpent, mostly due to its tiny head, flippers, and narrow body. In life, it may have grown to lengths of 70-80 feet, with the average individual measuring about 50 feet. While it shares many features with both archaeocetes and modern whales, *Basilosaurus* is characterized by its reduced hind legs, shortened pelvis, and lack of rear flippers, definite reminders of its terrestrial heritage. Its narrow, wedge-shaped, five-foot-long jaws held 44 teeth, among them well-developed incisors, canines, premolars, and multi-rooted molars. It was likely that these teeth were used to hold prey in the front of the jaw and help slice the food in the rear. The amount of wear on these teeth indicates that *Basilosaurus* likely ate fish and squid.

Zygorhiza, according to most scientists, is more similar to modern whales than is *Basilosaurus*, making it more advanced. However, in life *Zygorhiza* barely grew to a third of the length of *Basilosaurus*, reaching only twenty feet. *Zygorhiza* also had a relatively larger head and flippers than *Basilosaurus*, and possessed longer ribs. These features would have made it look less like a sea serpent than its larger, more primitive cousin. However, these two whales also share several characteristics. Both had complex, differentiated teeth, and both likely fed on fish and squid. It has even been hypothesized that both whales may have used their teeth not only to crush prey, but also as a strainer, to pick smaller prey from the water. Fossil evidence indicates that *Zygorhiza* and *Basilosaurus* had advanced ears, although these structures were internal.

The first known *Basilosaurus* material was discovered in 1832 by a wealthy Arkansas physician from the bluffs along the Ouachita River. This specimen, which consisted of over 28 vertebrae, was described in 1843 by local physician Dr. Richard Harlan, who thought he was describing the backbone of a large reptile. Because of his error, he named the animal *Basilosaurus cetoides*, which translates as "king of the lizards." Two years after the Arkansas discovery, in 1834, a complete skeleton of *Basilosaurus* was found on a plantation in southwestern Alabama. Staff members of the Philadelphia Academy of Natural Sciences, along with Sir Richard Owen, studied the skeleton and later concluded that it was not a lizard, but a carnivorous member of the whale family. Because of this, Owen renamed it "Zeuglodon," meaning "yoked-tooth." However, the rules that govern the naming of species recognize the earliest name given to an animal, even it if is inaccurate. Hence, the name *Basilosaurus* remains.

After the initial discoveries, specimens of *Basilosaurus* began turning up regularly in Louisiana, Mississippi, and Alabama. In 1845 a promoter, Albert Koch, excavated a specimen and exhibited it as a 114-foot-long sea serpent. While this was inaccurate, it turned the eyes of the scientific world towards the Gulf Coast area, where excavation was only catalyzed at a more feverish pace. During the late 1800s, the first well-preserved *Zygorhiza* specimens were discovered in Mississippi and Louisiana. Since that time an amazing *Basilosaurus* skeleton was unearthed in Madison County, Mississippi, and in 1971 a complete skeleton of *Zygorhiza* was excavated by paleontologists and members of the Mississippi Gem and Mineral Society near Tinsley.

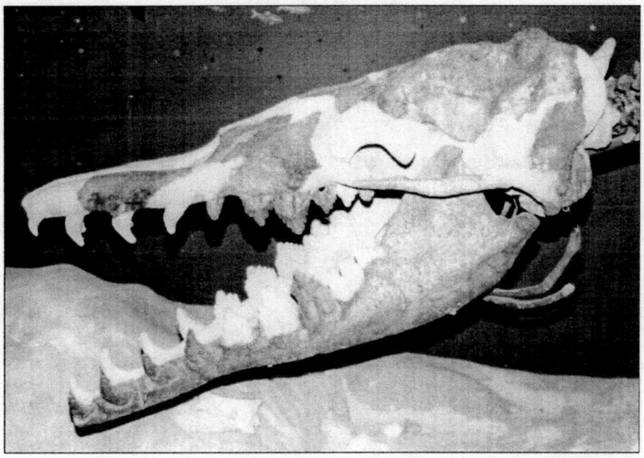

In the 1970s, the University of Southern Mississippi investigated two specimens in Jasper County, including one found by a farmer building a lake on his land. However, this excavation was overcome by greed and corruption, as the bulldozer operator hired to clear the site was found to have taken 23 vertebrae. After he promised to return them to the university, they mysteriously disappeared, and haven't turned up since. The story does have a happy ending, however—as bones from the discovery were used to assemble a composite specimen currently on display at the Smithsonian. Other archaeocete specimens can be seen at the Mississippi Museum of Natural Science in Jackson, the University of Southern Mississippi in Hattiesburg, and the Mississippi Office of Geology in Jackson.

Today, the "Basilosaur Belt" of Alabama, Mississippi, and Louisiana continues to yield fine specimens. These specimens are all found in sediments that make up the Jackson Group, including the Moodys Branch Formation and the Yazoo Clay. In fact, these specimens are so common that many local residents have been known to use whale vertebrae for fireplaces and blocks to support cabins.

On March 26, 1981, the "prehistoric whale," a designation referring to *Basilosaurus* and *Zygorhiza*, was designated the official Mississippi state fossil by the Mississippi Senate Concurrent Resolution No. 557. Though other groups of mammals may receive more publicity and attention from scientists, the whales make up one of the most intriguing and fascinating groups of animals to have ever lived. Much about their early evolution and history has been deduced from a plethora of fine fossils found in Mississippi.

FOR FURTHER INFORMATION:

Carpenter, K., and Dockery, D. T., 1985. "...And the bones came together, bone to bone." Ezekiel 37:7, The making of a state fossil. Mississippi Geology, vol. 6, no. 1., p. 1-6.

Carpenter, K., and White, D., 1986. Feeding in the archaeocete whale *Zygorhiza kochii* (Cetacea: Archaeoceti). Mississippi Geology, vol. 7, no. 2, pp. 1-14.

Daly, E., 1992. A List, Bibliography and Index of the fossil vertebrates of Mississippi. Office of Geology Bulletin, no. 128, Mississippi Dept. of Environmental Quality, Jackson, Mississippi.

Dockery, D. T., and Johnston, J. E., 1986. Excavation of an archaeocete whale, *Basilosaurus cetoides* (Owen), from Madison, Mississippi. Mississippi Geology, vol. 6, no. 3, p. 1-10.

Domning, D. P., 1969. A List, Bibliography, and Index of Fossil Vertebrates of Louisiana and Mississippi. Transactions of the Gulf Coast Association of Geological Societies, vol. 19, p. 385-422.

Frazier, M. K., 1980. Archaeocetes: whale-like mammals from the Eocene of Mississippi. Mississippi Geology, vol. 1, no. 2, p. 1-3.

Gingerich, P. D., Smith B. H., and Simons, E. L., 1990. Hind limbs of Eocene *Basilosaurus*: evidence of feet in whales. Science, vol. 249, p. 154-157.

Gingerich, P. D., Wells, N. A., Russell, D. E. and Ibrahim Shah, S. M., 1983. Origin of Whales in Epicontinental Remnant Seas. Science, vol. 220, pp. 403-406.

Heinrich, P.V., Primitive Eocene Whales, http://www.intersurf.com/~heinrich/Basilosaurus1.html

Johnston, J. E., 1991. Fossil Whale: State Fossil of Mississippi. Office of Geology Pamphlet no. 3, Mississippi Dept. of Environmental Quality, Jackson, Mississippi.

University of California at Berkeley, Cetaceans,
http://www.ucmp.berkeley.edu/mammal/cetacea/cetacean.html

MISSOURI

DELOCRINUS MISSOURIENSIS

Missouri has named a crinoid, a type of stalked echinoderm, as the official state fossil. The specific crinoid designated, *Delocrinus missouriensis*, is one of the most common fossils in the Show Me State, a large state known for both the antiquity and diversity of its fossil record. Although *Delocrinus* may not be the most flamboyant of state fossils, its ubiquity makes it a perfect fit for Missouri.

Since Missouri is famous for larger-than-life structures and personalities—the Gateway Arch, Mark Twain, and Harry Truman among them—it may be surprising that a simple crinoid is the state fossil. But crinoids may be the most abundant and widely-collected fossils in the world. Much of this abundance is due to the unique lifestyle and chemical composition of the crinoid, characteristics that make this strange animal treasured both by amateur and professional fossil collectors.

As mentioned, the crinoid is an animal, although its external appearance may lead one to believe it is anything but. Often dubbed a "sea lily," crinoids are members of the echinoderm lineage, a group of invertebrates closely related to the chordates, the phylum to which all vertebrates belong. According to crinoid paleontologist Charles Messing, the echinoderms make up one of the most fundamentally strange groups of animals known. The word echinoderm literally means "spiny skinned," a description fitting to the group. All echinoderms have a pentagonal symmetry, most obvious in the five arms of many sea stars and brittle stars, but also apparent in crinoids. All echinoderms also possess a unique internal system of water-filled vessels, much like blood vessels in humans, with side branches ending in small finger-like appendages called tube feet that may bear suckered tips. Once again, these tube feet are more obvious in starfish, but are also characteristic of crinoids. This water-filled network, technically known as the water vascular system, takes care of food capture, respiration, excretion, and also locomotion.

Echinoderms have a calcium carbonate skeleton, which includes interlocking plates in sea urchins and sand dollars, microscopic sclerites in sea cucumbers, and articulated ossicles in crinoids. These ossicles, which are nothing more than small circular plates, form the entire network of the crinoid's skeleton. Because of their tough chemical makeup, they preserve very easily, making them among the most common of fossils.

While over six thousand echinoderm species occur in the ocean, crinoids are the most unique. In his book *To the Deep Reef and Beyond*, Messing describes crinoids as the senior group of all living echinoderms, a moniker descriptive of their deeply-rooted family tree, which extends back for at least 500 million years.

Much of this long-standing pedigree can be traced to the crinoid's strange, but very successful, body form. Basically speaking, crinoids are separated into two groups: the sea lilies and feather stars. Although very similar, the sea lilies attach themselves to the ocean floor by a stalk, while the feather stars lack a stalk and are attached by a ring of hooks. Despite these differences, both groups of crinoids possess five arms, which often branch near their bases to give rise to more than ten smaller arms. These arms contain the aforementioned water vascular system,

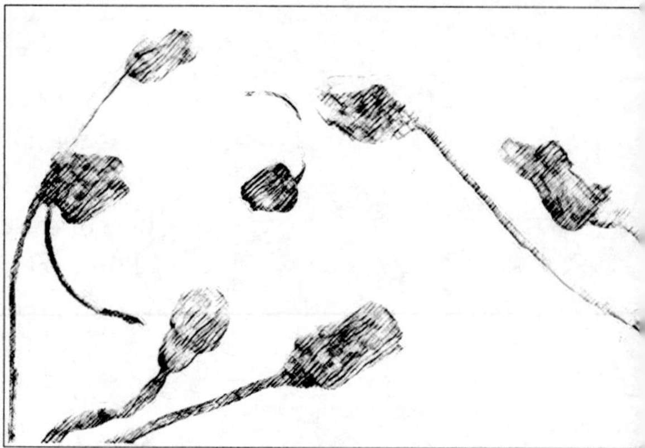

and end in tube-feet. Often these arms are found fossilized together, forming the crown of the crinoid. However, unlike the ossicles, complete crowns are quite rare in the fossil record.

Crinoids are filter feeders, making a living feeding off tiny species of plankton. Unlike most echinoderms, though, crinoids orient their mouths directly away from the substrate to which they attach. The arms surround these mouths, and are covered by slender side branches known as pinnules. These pinnules, which give the crinoid arm its feathery appearance, are lined with finger-like projections that flick passing food particles into a groove, where microscopic hairs carry them into the mouth.

Most of the crinoid body is composed of a basic skeleton of the aforementioned circular calcium carbonate ossicles, arranged in a series and articulated with ligaments and muscles. Because of this skeleton, some feather stars are able to crawl around, and some can even swim. This swimming ability is foreign to sea lilies, though. The crinoid's internal flesh is basically confined to a central mass containing the mouth, gut, and anus. Most echinoderms, crinoids included, lack a true heart and brain, but various nerves and blood vessels radiate through the arms.

The above information on crinoids is known as fact because a few crinoids are still living today. As a whole, their geological record extends from the Ordovician to recent, with a possible Cambrian specimen still giving paleontologists fits. Because of their unique skeleton, sea lily-type crinoids were one of the most common marine animals of the Paleozoic, occurring in dense clusters on the deep ocean floor, in tropical reefs, and even near the continental shelf. While they were common throughout the Paleozoic, crinoids were especially abundant during the Mississippian and Pennsylvanian, when they left great banks of debris, including a glut of ossicles. While crinoids nearly became extinct at the close of the Permian, they rebounded and continue to thrive today as "living fossils."

Delocrinus missouriensis lived about 280-300 million years ago, during the Pennsylvanian. During this time, Missouri looked much different than it does today. Obviously, it was covered by a sea. This warm, shallow sea surrounded the St. Francis Mountains, an ancient group of hills formed during the Cambrian that, while gone today, survived into the Pennsylvanian.

Pennsylvanian rocks, chock full of the ubiquitous crinoid ossicles, are well-known throughout Missouri, being especially common near Springfield and Kansas City. Pennsylvanian rocks also make up the Osage Plains, a section of the state comprised of rolling farmlands. Many of the crinoids found in Missouri are discovered in the Callaway and Burlington Formations. The Burlington Formation, specifically, outcrops near Springfield and Carthage, with King's Butte north of Springfield being one of the most popular locales. Sweeney Quarry near Clifton City in Cooper County is also known for its crinoidal limestone.

Because of the abundance of crinoid fossils in Paleozoic limestone, a group of Lee's Summit Elementary School students proposed the crinoid as the official state fossil in 1989. On June 16, 1989, the State Legislature rewarded their efforts by designating the crinoid, specifically writing in Act 10.090, "the fossilized remains of Crinoidea, scientifically designated as *Delocrinus missouriensis*, is declared to be the official state fossil of Missouri."

One might expect the Show Me State to designate a larger, showier state fossil. However, it is the ubiquitous crinoid that may be the most commonly collected fossil in Missouri. Because of its popularity with collectors and the effort of a dedicated group of students, the simple crinoid bears this lauded designation.

FOR FURTHER INFORMATION:

Cowen, R., 2000. The History of Life. Blackwell Science, London, 432 p.

McMillen, M., Missouri's Massive Mammals, http://www.conservation.state.mo.us/kids/out-in/1997/4/1.html

Messing, C.G., 1987. To The Deep Reef and Beyond. Deep Ocean Society, Miami, 30 p.

Unklesbury, A.G., 1955. Common Fossils of Missouri. University of Missouri Handbook Number 4.

Missouri Secretary of State, Missouri State Fossil, http://mosl.sos.state.mo.us/ofman/syfosl.html

MONTANA

MAIASAURA

One of the most important dinosaurs discovered in the last quarter century was found near a small Montana town. This discovery, publicized by now-famous paleontologist Jack Horner, helped prove that dinosaurs were not stupid, slow-moving creatures, but instead active, intelligent animals with social instincts. This dinosaur, *Maiasaura peeblesorum*, is so important that it has been designated Montana's official state fossil.

Maiasaura lived during the Late Cretaceous, about 77 to 73 million years ago. During this time, the Rocky Mountains were just beginning to form, and Montana was a lush flood plain bordering the Western Interior Seaway, which stretched from the Gulf of Mexico to the Arctic Ocean. *Maiasaura* was a member of the hadrosaur lineage, a subfamily of diverse duck-billed dinosaurs which were common during the Late Cretaceous in North America. Like other hadrosaurs, *Maiasaura* had a long, wide face with a short, wide bill. Above and between its eyes was a small, solid crest, perhaps used as a display structure to impress mates, or as a resonating chamber for sound. The skull of *Maiasaura* was quite large, measuring somewhere around 810-830 millimeters (about 32 inches) long, and about 350 millimeters (13.5 inches) high. In addition to being large, this skull also shows an interesting mix of primitive and advanced characteristics, including the small nostril opening and the broad nasal bones at the front end of the muzzle.

Maiasaura, like all hadrosaurs, was an herbivore, and used its long, toothless beak, probable check pouches, and self-sharpening cheek teeth to become an efficient plant-eater. Of these features, the teeth are the most interesting, as they were covered by hard enamel on both the outer surface of the upper teeth and the inner surface of the lower teeth. It is also thought that *Maiasaura* manipulated plant material with its hands, which contained four fingers.

Maiasaura was a large animal, which grew to lengths of more than 9 meters (30 feet), and likely weighed 3-4 tons. *Maiasaura* usually walked quadrupedally, but also may have occasionally reared up on its hind legs to reach higher vegetation. The front legs of this dinosaur were shorter than its rear legs, which means that running was also probably accomplished bipedally.

The significance of *Maiasaura* is not in its size, primitive characteristics, or running ability, however. Instead, this was the first dinosaur to have been found alongside its young, eggs, and nests. Based on these discoveries, it has been hypothesized that *Maiasaura* cared for its young—hence its name, which means "good mother lizard."

The discovery of *Maiasaura* and its nests, babies, and bone beds can be traced back to 1977, when a young Jack Horner boasted to friends that he would one day find dinosaur eggs and baby skeletons in Montana. Something drew Horner to the Two Medicine Formation of the Big Sky State, and something told him that babies could, and would, be found. In order to fulfill his brag, Horner contacted his old boyhood friend and fellow dinosaur maven, Bob Makela, and asked him to help locate the elusive fossils. Makela, a high school teacher, agreed, and before long Horner and Makela found themselves in John and Marion Brandvold's T. rex Agate Shop, a rock shop located in a former church, in the town of Bynum, Montana. There the Brandvolds showed Horner a coffee can full of small bones they had discovered near the town of Choteau. Horner immediately recognized what he had in front of him: the bones of baby dinosaurs he had predicted a year earlier. The year was 1978, and Horner's prognostication had come true!

The Brandvolds agreed to give Horner the bones and show him where they had found them. They took the young paleontologist and his friend to the

Peebles' Ranch near Choteau, where before long Horner had uncovered the first dinosaur nest. After some prospecting, Horner decided to study what he had found and return the next year. Upon their return, Horner and Makela found not only several more nests, but bones of nestlings, juveniles, and adults all in the same stratum—evidence that indicated to him that these animals were living together. In addition, the bones of the young maiasaurs were chock full of tiny blood vessels, which indicated possible warm-bloodedness. Later that field season, Horner discovered perhaps the most famous paleontological site of the late 20th century: Egg Mountain. Named for its plethora of eggs, this site continues to yield fine specimens today. Based on their discoveries, Horner and Makela named their new dinosaur *Maiasaura peeblesorum*, honoring both its motherly instincts and the Peebles, who owned the ranch. The Egg Mountain site remained in the Peebles' possession until 1987, when they sold it to the Nature Conservancy, which leads annual digs there.

The nests of *Maiasaura* were basically holes scooped out of the ground, usually about 2 meters (6-7 feet) in diameter. Approximately 15-30 grapefruit-sized eggs could fit in such a nest. Many of these nests were found preserved along with hatchlings, juveniles, and adults. Newborns were about a third of a meter (1 foot) long, while babies were about 1 meter (3.3 feet) long. Horner reported the discovery of four babies no more than two meters (7 feet) from a nest, suggesting that they stayed close to their place of birth and were cared for

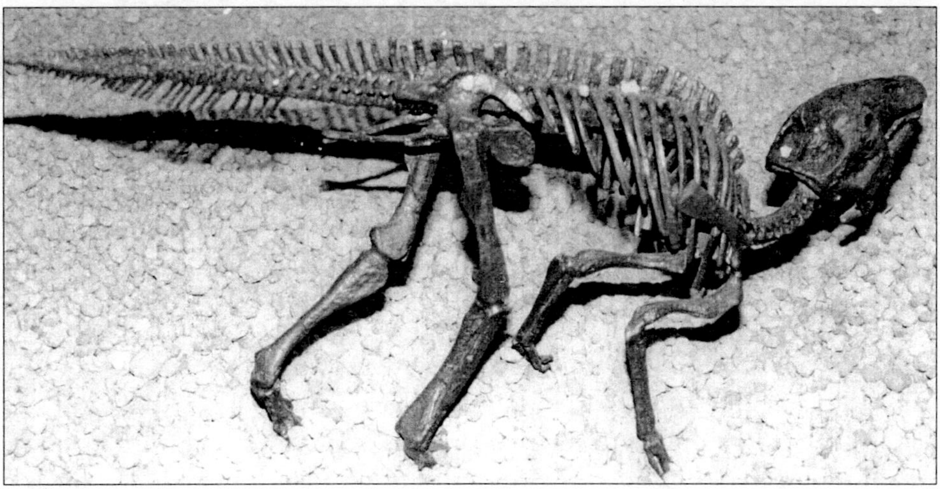

by parents. These juveniles had short snouts and relatively large feet, but otherwise resembled the adult maiasaurs.

Because there was such a size disparity between the hatchlings and the other youngsters found in the vicinity, and because of the aforementioned blood vessel-rich bone, Horner suggested that *Maiasaura* grew relatively quickly, possibly indicating warm-bloodedness, at least early in life. He also hypothesized that *Maiasaura* was a gregarious animal, and annually congregated at communal nesting sites to give birth to and care for its young. Nests at one of these sites number over 40, and cover some 2.5 acres of land. As a general rule, Horner found that nests were placed 25-30 feet apart, about the size of an adult *Maiasaura*, meaning that each mother was provided ample room to lay her eggs. In addition, trampled eggshells suggested to Horner that the hatchlings stayed in the nests for a prolonged period of time.

Based on this evidence, Horner and his colleagues formulated a hypothesis, stating that *Maiasaura* may have nested in large colonies and returned to the same breeding grounds each year. During the dry season, perhaps, the mother made her nest on the floodplain, using fine-grained silts and sands, and then laid her eggs in a spiral pattern. The nests were covered with vegetation, which provided insulation and even generated some heat as it rotted. (The modern Nile crocodile is known to do this.) After the babies hatched, the colony would have stayed put until the youngsters were old enough to move on. Such a hypothesis is, of necessity, somewhat speculative in nature—but in keeping with the fossil evidence. Another location where some 10,000 *Maiasaura* individuals are preserved also lends credence to this scenario. Horner believes that the animals composed a herd, and were drowned in a flash flood.

Because of its obvious scientific importance, *Maiasaura* was honored as the official state fossil in 1985. Ben Veach, former president of the Montana State Council of Rocks and Minerals, led the drive to crown *Maiasaura*. His enthusiasm garnered the support of Livingston Middle School sixth-grade teacher Helen Peterson. Peterson urged her students to write letters to the governor and legislators, and to send petitions to other schools across the state. Her efforts rubbed off on Representative Orval Ellison of McLeod, who sponsored a bill to designate *Maiasaura*. After Ellison introduced the bill, hundreds of students lobbied in the Montana capital building, even bringing 2,000 dinosaur cookies to give to the lawmakers. The result was a unanimous "yea" from the legislature. Later, Bill 130 was signed into law by Governor Ted Schwinden, on February 22, 1985.

Today, fossil fans longing to see *Maiasaura* can check out fossils at the Museum of the Rockies in Bozeman, Montana, the Old Trail Museum in Choteau, and the National Museum of Natural History in Washington, D.C. Adventurous dinosaur buffs can even join a dig at the Egg Mountain site.

While arid and sparsely populated today, Cretaceous Montana was lush, full of vegetation, and home to a plethora of dinosaurs. One of these, *Maiasaura*, taught science that dinosaurs were not sluggish, stupid creatures, but rather fast-moving, social animals. This dinosaur remains one of the most important paleontological discoveries of the past quarter century, and is more than worthy of state fossil designation.

FOR FURTHER INFORMATION:

Brusatte, S., 2001. Kids' Corner-*Maiasaura*. Fossil News: Journal of Avocational Paleontology, vol. 7, no. 3, p. 12.

Carpenter, K., Hirsch, K.F., and Horner, J.R. (eds.), 1994. Dinosaur Eggs and Babies. Cambridge University Press, New York.

Carpenter, K., 1999. Eggs, Nests, and Baby Dinosaurs. Indiana University Press, Bloomington, 336 p.

Holmes, T., and Holmes, L., 2001. Peaceful Plant Eating Dinosaurs. Enslow Publishers, Berkeley Heights, New Jersey, 128 p.

Horner, J.R., and Makela, R., 1979. Nest of Juveniles Provides Evidence of Family Structure Among Dinosaurs. Nature, v. 82, p. 296-298.

Horner, J.R., 1982. Evidence of Colony Nesting and "Site Fidelity" Among Ornithischian Dinosaurs. Nature, v. 297, p. 675-676.

Horner, J.R., and Gorman, J., 1988. Digging Dinosaurs. Workman Publishing, New York, 210 p.

Horner, J.R., and Dobb, E., 1997. Dinosaur Lives: Unearthing an Evolutionary Saga. Harcourt Brace and Company, New York, 244 p.

Lessem, D., 1992. Dinosaurs Rediscovered. Simon and Schuster, New York, 368 p.

NEBRASKA

MAMMOTH

Unlike many other states, Nebraska looks nearly the same today as it did during the Ice Age. However, in place of endless fields of wheat and corn were found vast grasslands and highlands, inhabited by a unique and intriguing fauna. One of the key members of this fauna, the mammoth, has been designated the official state fossil of Nebraska.

Since little of Nebraska was covered by glaciers, vast prairies were able to develop during the Pleistocene. These grasslands supported a far greater diversity of large mammals than most land habitats do today. Among other animals, mountain sheep inhabited the higher portions of these plains, while vultures larger than today's California condor soared overhead, searching for dead bison, elephants, and other large animals. Vast flocks of waterfowl fed and nested near small lakes. These lakes, several of which remain today, were created when the few glaciers that reached Nebraska retreated. Near these lakes lived small mammals such as beavers and prairie dogs, and large beasts like the state fossil, the mammoth.

Three distinct species of mammoth are known from Nebraska: the woolly mammoth *(Mammuthus primigenius)*, the Columbian mammoth *(Mammuthus columbi)*, and the Imperial mammoth *(Mammuthus imperator)*. Taken together, mammoths are members of the Family Elephantidae, a group of mammals which

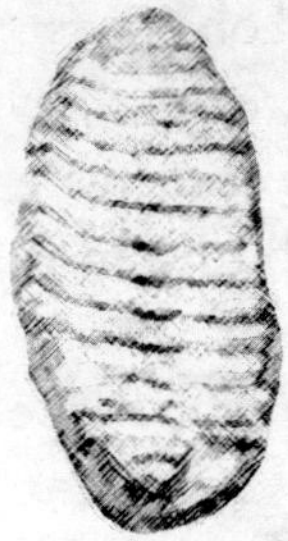

also includes the extinct gomphotheres, as well as modern elephants. Generally, mammoths were grazers, feeding mainly on grass and flowering plants. The best scientific estimates say that a mammoth individual probably had a lifespan of 70-80 years. During this time, the mammoth would go through up to six sets of flat molar teeth. The largest set of these teeth, which are typically found fossilized, were about the size of a large shoe box and had a flat, ridged surface adapted for grinding up tough plant material. They are highly recognizable as fossils and are sometimes described as looking like the soles of tennis shoes or a stack of Fig Newtons.

Mammuthus imperator, which is known from Nebraska, was the largest of the mammoth species, and grew to sizes of well over 4 meters (13 feet). All mammoths possessed wide heads, a sloping back, and two long, curved tusks. Depending on the species, these tusks, which are also commonly found fossilized, could reach several feet in length. According to most scientists, these tusks were used to both help gather food and to wrestle with other members of the species. Studies of modern elephants have shown that tusks can be efficiently used to gather plant material from on or near the ground, and are also often used by males to wrestle. This wrestling, which was likely an activity used to impress potential mates, would have been accomplished by interlocking the tusks. According to one interesting study, the inner surface of one tusk tends to be more worn than the other, indicating that some mammoths may have been "right-tusked" or "left-tusked" in their food-gathering behavior—like modern humans are right- or left-handed.

Mammoths evolved in Eurasia, and entered North America via the Bering Land Bridge—when Alaska and Siberia were connected during a time of lowered sea level. During the Pleistocene Epoch, which lasted from about 1.6 million to 10,000 years ago, mammoths were commonly found in grassland habitats throughout the Northern Hemisphere. Today frozen mammoths can still be found in Siberia, a remnant of the icy past. However, one of the best places to find fossilized mammoths is in Nebraska.

Mammoth remains have been reported from all 93 of Nebraska's counties, and the University of Nebraska alone has catalogued bones found in 84

counties. Based on one scientific estimate, the remains of no fewer than ten mammoths lie buried in an average square mile of Nebraska landscape! These mammoths are so common that bones and tusks have been found just one block from the University of Nebraska State Museum's famous Elephant Hall, and mammoth remains were even uncovered when the museum was digging a basement office for a staff paleontologist!

The most famous mammoth discovery in Nebraska took place in Lincoln County in 1922. In that year, a nearly-complete, beautiful, and gigantic specimen of the Imperial mammoth was discovered. Nicknamed "Archie" because of its scientific name, *Archidiskodon (Mammuthus) imperator maibeni*, this beast stands 14 feet high at the shoulder—and it was not fully grown at the time of its death. Scientific estimates place its probable weight at about 15 tons. Largely studied by University of Nebraska State Museum director E.H. Barbour, this specimen was mounted in the Museum's Elephant Hall in 1933. On October 21, 1998 a nearly life-size bronze statue of Archie was dedicated in front of Morrill Hall on the University's campus. This statue is only 12 feet tall at the shoulder, but is impressive nonetheless.

Good mammoth remains are known from many other counties, including Dawes, Sioux, and Sheridan counties. Other than the Lincoln County specimen, the most famous mammoth discovery in Nebraska took place in 1867, when U.S. Government surveyor and naturalist Ferdinand Vandiveer Hayden discovered a cache of mammoth bones in Sheridan County, south of Rushville and Hay Springs. The best mammoth discoveries are usually found in the western part of Nebraska, and often occur along or near the state's major rivers, among them the Missouri, Platte, Niobrara, Blue, Republican, and Nemaha.

Fossil mammoth bones in Nebraska have also shown evidence of human interaction. At a few isolated sites, spear tips and other evidence of human hunting have been found alongside mammoth and other Pleistocene mammal bones, suggesting that the animals were killed by early human hunters. In fact, human predation is one major candidate for the extinction of the mammoth, which disappeared from North America about 11,000-10,000 years ago. According to a recent study by University of California at Santa Barbara biologist John Alroy, it is more than plausible that human hunters entirely wiped out the mammoth and other large Ice Age mammals. However, American Museum of Natural History paleontologist Ross MacPhee believes that humans didn't cause the extinction of the mammoth by overhunting, but rather by spreading a new, deadly disease. Regardless, it is possible that humans did play a hand in one of the most devastating and puzzling extinctions known.

Despite mankind's possible role in the extinction of the mammoth, humans have long been fascinated by this animal. Its remains have intrigued both Thomas Jefferson and Georges Cuvier, and its mounted skeletons are common in museums around the world. This holds true in Nebraska, where, in addition to Archie at the University of Nebraska State Museum, a near-complete mammoth discovered near Crawford is on display at the Trailside Museum in Fort Robinson State Park. The popularity of these specimens, the ubiquity of their discovery, and their scientific importance led the Nebraska State Legislature to name the "mammoth" the official state fossil on March 1, 1967, which was, coincidentally, the 100th anniversary of the state.

While prehistoric humans may have helped kill them, people today are fascinated by mammoths. Luckily for science, the state of Nebraska has been more than kind, yielding a plethora of mammoth specimens. Although the state fossil was named over 30 years ago, good mammoth fossils continue to turn up today, and the lode shows no signs of running dry.

FOR FURTHER INFORMATION:

Barbour, E.H., 1913. A Tooth of the Columbian Mammoth. Nebraska Geological Survey Bulletin Number 4-5, 7 p.

—, 1925. *Elephas scotti*, a New Primitive Mammoth from Nebraska. Nebraska Geological Survey Bulletin Number B-1-2, 4 p.

Brusatte, S., 2001. Kids' Corner-Mammoth. Fossil News: Journal of Avocational Paleontology, vol. 7, no. 4, p. 11.

Cook, H.J., 1914. Note on the Occurrence of the Mammoth in Sioux County, Nebraska. Nebraska Geological Survey Bulletin Number 7-6, 3 p.

Lugn, A.L., Outline of Pleistocene Geology of Nebraska, in: The Geology and Mammalian Fauna of the Pleistocene of Nebraska, University of Nebraska State Museum Bulletin Number B-1-41, 38 p.

Schultz, C.B., Pleistocene Mammals of Nebraska, in: The Geology and Mammalian Fauna of the Pleistocene of Nebraska, University of Nebraska State Museum Bulletin Number B-1-41, 38 p.

Nebraska Blue Book, 1998-1999, p. 15.

University of Nebraska State Museum, Vertebrate Paleontology, http://www.museum.unl.edu/research/vertpaleo/vertpaleo.html

University of Nebraska-Lincoln, Mammoth, http://csd.unl.edu/csd/trivia/statestuff/fossil.html

NEVADA

SHONISAURUS POPULARIS

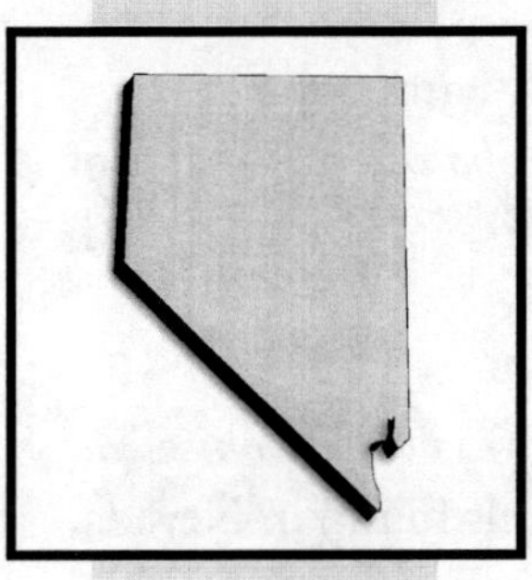

Today, Nevada lies almost entirely within the Great Basin, an area between the Rockies and the Sierra Nevada Mountains that drains to no ocean. During the Late Triassic, Nevada also lay within a large basin, one filled with a vast system of shallow seas and lakes. Populating these seas were cephalopods, nautiloids, fish, and several intriguing and mysterious ichthyosaurs. One ichthyosaur, the 225-million-year-old *Shonisaurus popularis*, has been designated the state fossil.

Like all other ichthyosaurs, *Shonisaurus* looked much like an oversized dolphin, with a sharp snout, large eyes, and streamlined head and body. Its dorsal fin was high, its tail was vertical, and its limbs were modified into large flippers. These features were all adaptations for controlled, fast swimming. All ichthyosaurs had four flippers, used for propulsion and stability, which ranged from large and stocky to quite small. Over time, the vertical tail evolved a sharp, downward bend, which forced the rear end of the individual downward while swimming, an action countered by the four flippers. The vertebrae of ichthyosaurs were shaped like flattened discs, which enabled the animal to swim with a fast, undulating movement. Most ichthyosaur jaws were studded with rows of conical teeth, which sometimes reached 200 in number. These teeth, along with the versatility of the ichthyosaur in water, made it a successful predator. Most scientists believe that ichthyosaurs ate fish and cephalopods, a hypothesis which has been strengthened by discoveries of ammonites and other molluscs alongside ichthyosaurs in Nevada.

Shonisaurus is the largest ichthyosaur known from well-preserved fossil skeletons. Based on evidence found in Nevada, it has been estimated that *Shonisaurus popularis* grew to lengths of 45-50 feet, possibly even longer, and may have weighed 40 tons. This size is comparable to that of the modern sperm whale. In addition, larger *Shonisaurus* individuals had six-foot-long front fins, twenty-five-foot-long tails, and a ten-foot-long skull with elongate jaws, filled with the aforementioned conical teeth. Although *Shonisaurus* may not have been as good a swimmer as some other ichthyosaurs, its size and teeth assured it a successful predatory niche.

As a group, ichthyosaurs first appeared in the Triassic, making *Shonisaurus* a valuable early genus to study. Largely because of the fossils found in Nevada, scientists have been able to construct a typical Late Triassic marine environment, with the ichthyosaur playing the role of top predator. Ichthyosaurs continued to evolve more streamlined shapes in the Jurassic Period, and eventually became the undisputed kings of the sea. However, near the end of the Cretaceous, all ichthyosaurs became extinct. The cause for this is unknown, but most scientists believe that it is unrelated to the extinction of the dinosaurs.

Regardless, *Shonisaurus* went extinct long before the ichthyosaur lineage died out. Fortunately, it left some wonderfully-preserved fossils in Nevada. Most of these fossils are found in a remote area in the Shoshone Mountains, now dubbed Berlin-Ichthyosaur State Park. This state park lies close to the ghost town of Berlin, a turn-of-the-century mining camp that employed over 250 silver and gold miners. These miners were the first to find *Shonisaurus* fossils, and reportedly used them as dinner plates and decorations in their fireplace hearths.

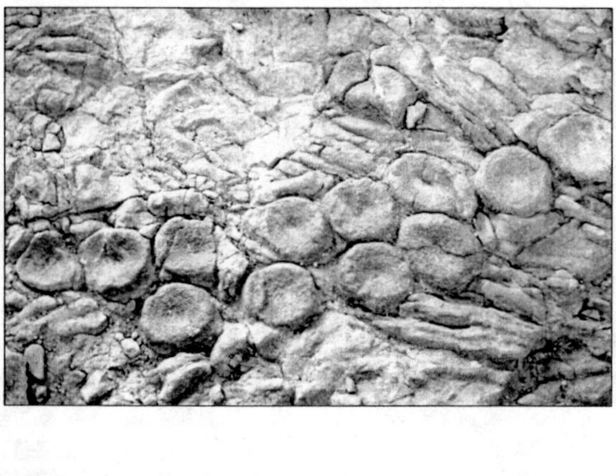

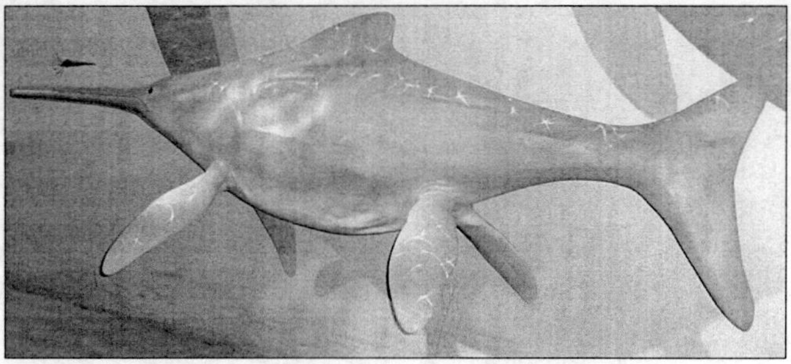

In 1928 these miners also became the first to find relatively complete specimens of *Shonisaurus*. While returning from work one afternoon, a group of miners noticed a set of strange skeletons eroding from a sandstone cliff. Soon the site was brought to the attention of scientists, who determined that 37 ichthyosaur individuals lay preserved in the rock. However, it was not until nearly 30 years later, in 1954, that a full-scale excavation was launched by University of California-Berkeley paleontologists Charles L. Camp and Samuel P. Welles. While in Berlin, Camp described three ichthyosaur species, including *Shonisaurus popularis*. Seeing the importance of the Berlin site, the Nevada state government designated the area Ichthyosaur Paleontological State Monument in 1955, a mere year after the excavation began. A decade later, in 1966, an A-frame shelter was built over the fossil site, leaving in place the remains of nine ichthyosaurs. Finally, four years later, in 1970, the mining claims of Berlin were purchased, and the park was renamed Berlin-Ichthyosaur State Park.

The fossils of the *Shonisaurus* individuals found at Berlin reveal a great deal about the Triassic environment of Nevada. At the time *Shonisaurus* lived, Nevada was situated along the west coast of what is now North America. Sediments from the Luning Formation, the rock unit in which *Shonisaurus* was discovered, indicate that the ichthyosaur bones were deposited in a deep ocean shelf environment. Fossils found alongside *Shonisaurus* also support this hypothesis.

However, it is not well understood exactly how 37 ichthyosaurs died and became preserved in one small area. Originally, Camp devised a hypothesis in which the ichthyosaurs became confused and beached themselves on mudflats, much as modern whales sometimes do today. This hypothesis was supported by the fact that all of the Berlin specimens were adults, and most of them were lying in a generally north-south direction. However, this hypothesis has come under fire in recent years. Today, most scientists believe that the individuals simply died, sank to the sea floor, and their carcasses were deposited in a shallow groove by weak ocean currents. Scientists have also devised a possible explanation for their deaths: food poisoning. It has been suggested that the ichthyosaurs ate fish or shellfish tainted with a neurotoxin that paralyzed them. This food poisoning is sometimes seen in modern whale communities off the coast of New England.

The hypothesis of *Shonisaurus*' death isn't the only revised notion regarding this ichthyosaur. During the past decade much of the anatomy of this animal has also been restudied. When Camp originally published his description of *Shonisaurus* in 1976, he included with the paper an erroneous reconstruction of the ichthyosaur, a drawing that has been commonly reproduced and copied by various paleoartists. In 1990 Bradley Kosch, a paleontologist at Lake Tahoe Nevada State Park, published a reexamination of *Shonisaurus*, and concluded that Camp's illustration was all wrong. Among other problems, Kosch noted that in Camp's reconstruction the skull, ribs, and paddles were too large, the tail was too long, and the animal was portrayed with too wide a body. A bas relief of the large-bodied restoration adorns the concrete exterior of the visitor's center. Unfortunately, few paleoartists have followed Kosch's reexamination. A notable exception is Shannon Yeager, who first published her "new" *Shonisaurus* reconstruction (right) in the June, 2001 issue of *Fossil News*.

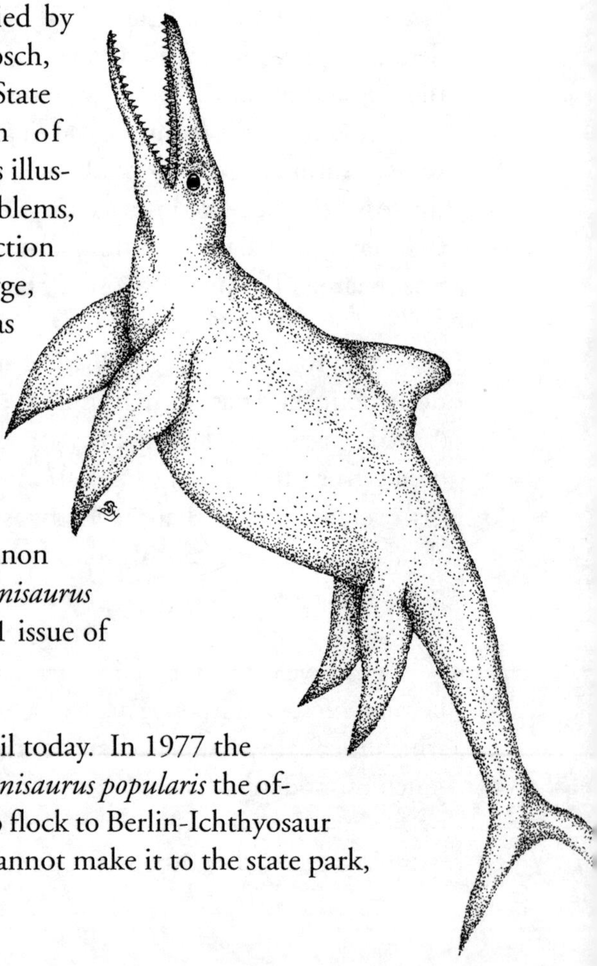

Shonisaurus remains a popular fossil today. In 1977 the Nevada State Legislature designated *Shonisaurus popularis* the official state fossil, and visitors continue to flock to Berlin-Ichthyosaur State Park year round. For those who cannot make it to the state park, remains of *Shonisaurus* are also displayed in two Nevada museums: the Las Vegas Museum of Natural History and the Nevada State Museum. In fact, the Las Vegas Museum recently acquired the type specimen of *Shonisaurus*, the famous skeleton discovered by the miners in 1928. Previously, this specimen had been exhibited at the Barrick Museum at UNLV. Its new location makes it more accessible to scientists and the public.

Although it is sparsely populated today, Nevada can claim a unique form of revenue in casino gambling, and an even more unique state fossil, the ichthyosaur *Shonisaurus popularis.* Although it lived in shallow seas some 225 million years ago, remains of this creature are found today 7,000 feet above sea level near the ghost town of Berlin. Luckily for science, the miners of Berlin had more success discovering fossils than finding gold.

FOR FURTHER INFORMATION:

Brusatte, S., 2000. Kids' Corner: *Shonisaurus popularis.* Fossil News: Journal of Avocational Paleontology, vol. 6, no. 12, p. 11.

Camp, C.L., 1976. Large Ichthyosaurs from the upper Triassic of Nevada. Palaeontographica 170.

Debus, A.A. and Debus, D., 2001a. The Big Sho- (Part 1 of 2). Fossil News: Journal of Avocational Paleontology, vol. 7, no. 5, p. 5-10.

—, 2001b. The Big Sho- (Part 2 of 2). Fossil News: Journal of Avocational Paleontology, vol. 7, no. 6, p. 14-18.

Kosch, B., 1990. Large Ichthyosaurs from the Upper Triassic of Nevada. Journal of Vertebrate Paleontology, vol. 10, no.4, p. 512-514.

Orndorff, R.L., Wieder, R.W., and Filkorn, H.F., 2001a. How the West Was Swum. Natural History, vol. 110, no. 5, p. 22-24.

—, 2001b. Geology Underfoot in Central Nevada. Mountain Press Publishing Company, Missoula, Montana.

Berlin-Ichthyosaur State Park
HC 61, Box 61200, Austin, NV 89310; Phone: 702-964-2440
Hours: Fossil site tours 10 a.m., 2 p.m. and 4 p.m., Memorial Day-Labor Day; weekends, daily, 10 a.m. and 2 p.m., after Labor Day; by reservation in winter.

NEW MEXICO

COELOPHYSIS

The Late Triassic world was much different than what we imagine as the age of dinosaurs. All of the continents were gathered together in a single land mass known as Pangaea. Except for small islands, the remainder of the globe was occupied by a deep ocean. This single land mass and vast ocean created strange weather patterns, and often led to prolonged seasons of rain and drought (monsoons). On this supercontinent arose one of the most successful groups to have ever lived—the dinosaurs. One of the earliest well-known dinosaurs, *Coelophysis*, is designated as New Mexico's state fossil.

Coelophysis lived during the Late Triassic, about 210 million years ago. This theropod, which has been found in New Mexico, Utah, and Arizona, lived in what was likely a seasonally dry, desert-like environment, possibly comparable to a savanna-type climate without grass. (Grasses and true savannas did not become widespread until the Miocene.) However, seasonal rains may have annually interrupted the xeric conditions, creating a short period of humidity and lush vegetation growth. These rains may have also created harsh floods, which not only robbed the area of its vegetation, but also may have led to the sudden death of numerous *Coelophysis* individuals, a topic we shall visit in a moment.

Based on its common fossil remains, *Coelophysis* was quite a successful dinosaur. Although sizes vary, the average *Coelophysis* individual measured anywhere from 2.5-3.5 meters (about 8-10 feet) long, and weighed somewhere from 50-150 pounds. *Coelophysis* was a small, slender, lightweight theropod. It had a long, flexible neck and tail, and walked upright on its strong, yet slender, hind limbs. Its front limbs ended in three clawed fingers that were used to grasp and tear prey. The head of *Coelophysis,* like its body, was small and strong, with sturdy jaws that contained about 100 teeth.

These teeth, along with a host of other characteristics, made *Coelophysis* a successful predator. Many scientists have suggested that the dinosaur may have eaten small lizards, amphibians, and even other dinosaurs, possibly heterodontosaurs. Some even believe it to have been a scavenger, picking at the remains of animals killed by other dinosaurs. However, based on a discovery in New Mexico, it is now thought that *Coelophysis* may have also been a cannibal. In 1947-1948, a crew from the American Museum of Natural History in New York excavated dozens of *Coelophysis* remains near Abiquiu, New Mexico, in the northern part of the state. We shall discuss the history behind these discoveries in a moment. What is amazing, though, is what was found inside the abdominal cavities of several individuals: the bones of young *Coelophysis* individuals. These finds initially caused scientists to hypothesize that *Coelophysis* gave birth to live young. But, the pelvic opening was too small to permit a live birth. Instead, some think that these individuals were cannibals, and ate the young of their own species, perhaps during times of drought or starvation (although others believe this to be a preservational artifact).

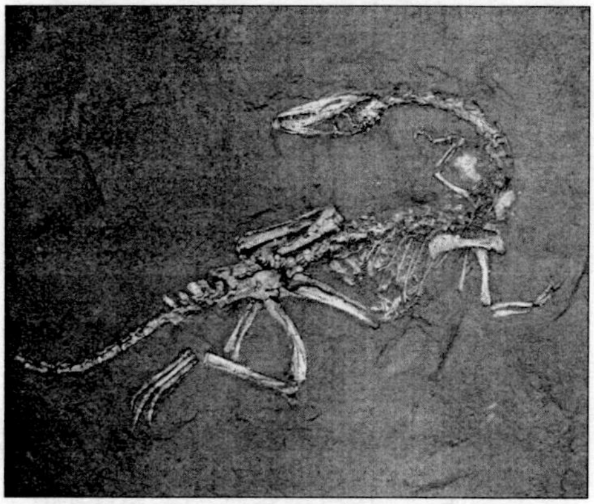

The Abiquiu finds have also given new insight into theropod sexual dimorphism. While paleontologists working at the site were studying several of the *Coelophysis* remains, they came across two odd specimens. These specimens, now housed in the collections of the AMNH, exhibited different characteristics in the skull, neck, forelimbs, and sacrum. Perhaps, paleontologist Edwin

Colbert said, the skeleton with the larger skull, longer neck, smaller forelimbs, and fused sacral spines is a male. This judgment may be erroneous, since it is based on the idea that a larger skull and longer neck were advantageous in predation and combat. The real answer may never be known.

As with cannibalism and sexual dimorphism, there is also argument over the proper taxonomic name and position of *Coelophysis*. This dinosaur was originally named in the 1800s based on fragmentary and isolated material. However, much new information has come to light after the AMNH discoveries. But one group of scientists argued that the AMNH specimens may not be the same as the original *Coelophysis* specimens discovered in the 1800s—therefore, they proposed a new name for the dinosaur: *Rioarribasaurus*. Later, the International Commission on Zoological Nomenclature voted to redefine the type specimen of *Coelophysis* as one of the AMNH specimens, so *Rioarribasaurus* is not valid. The original *Coelophysis* material, which may or may not belong to the same animal as the AMNH specimens, has been given its own dubious name: *Eucoelophysis*.

Much of this confusion dates back to 1881, when fossil hunter David Baldwin discovered the first *Coelophysis* bones in northwestern New Mexico. Baldwin, once a trusted O.C. Marsh assistant, later began to work with Edward Cope, Marsh's scientific rival. A peculiar scientist, Baldwin would often excavate fossils in the dead of winter, accompanied only by his mule. During the winter of 1881 he came across the first small bones of *Coelophysis*, and sent

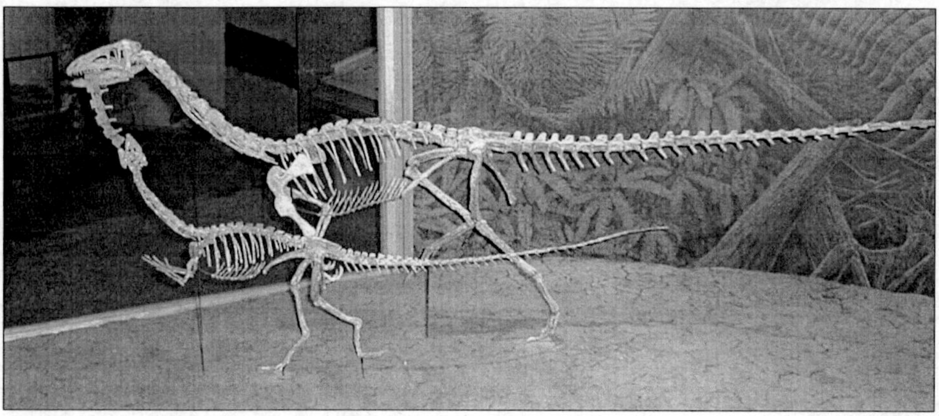

them to Cope. In 1889 Cope named the bones *Coelophysis*, and created three species. Today, only *Coelophysis bauri* stands as a valid species name.

For over 60 years, *Coelophysis* remained a poorly-understood dinosaur. That all changed in 1947, when the aforementioned expedition from the American Museum discovered a concentration of *Coelophysis* bones on Ghost Ranch near Abiquiu, close to where Baldwin had discovered his fossils in the late 1800s. The first bones discovered by the expedition were found unceremoniously, when museum paleontologist George Whitaker followed a trail of fossil scraps up a crumbling hillside to a ledge of bone eroding out of the hill. He brought a few samples to fellow AMNH paleontologist Edwin Colbert, who followed Whitaker back to the site and discovered a small claw. Thus began a large excavation at the site. To date, this excavation has turned out dozens of complete *Coelophysis* skeletons, including a plethora of skulls and the previously-mentioned "cannibal" specimens. A small but well-done museum, the Ruth Hall Museum of Paleontology, is open to the public at the Ghost Ranch site and exhibits many *Coelophysis* specimens, as well as other fossils.

After the AMNH dig ended, all scientific expeditions were stopped at Ghost Ranch, until the Carnegie Museum of Natural History began further excavations in 1981. One of the Carnegie's discoveries, a perfectly-preserved skull, became the first dinosaur fossil to travel into space, when it took off on the Endeavor Space Shuttle and was transferred to the Mir Space Station on January 22, 1998. Since the Carnegie began its excavation, several new specimens have been discovered, mounted, and exhibited, including those at the Carnegie Museum, American Museum, and Ruth Hall Museum. Since that time it has also been hypothesized that the cache of dinosaur bones discovered at Ghost Ranch represents a herd which was suddenly drowned in a flash flood, a notion which is convincing but hard to prove.

It was not until 1981, however, that *Coelophysis* became New Mexico's official state fossil. The credit for this designation goes to the New Mexico Museum of Natural History, which came up with the idea of adopting a state fossil to help raise money to build the museum. Two dinosaurs were considered, *Pentaceratops* and *Coelophysis*, and on March 17, 1981 the latter was named the official state fossil.

Although it was originally discovered as fragmentary and isolated fossils, *Coelophysis* has become one of the better-known Triassic dinosaurs after the discovery of an amazing bonebed in northern New Mexico. Thanks to the efforts of the New Mexico Museum of Natural History and the State Legislature, this small, slender theropod is now one of the most popular Triassic dinosaurs with schoolchildren and scientists alike.

FOR FURTHER INFORMATION:

Brusatte, S., 2001. Kids' Corner-*Coelophysis*. Fossil News: Journal of Avocational Paleontology, v. 7, no. 8, p.10.

Colbert, E.H., 1947. The Little Dinosaurs of Ghost Ranch. Natural History, 61:393-399, 427-428.

—, 1960. Triassic Rocks and Fossils. Guidebook, New Mexico Geological Society, 11th Field Conference: 55-62.

—, 1968. The Great Dinosaur Hunters and Their Discoveries. E.P. Dutton & Company Publishers, New York.

—, 1974. the Triassic Paleontology of Ghost Ranch. New Mexico Geological Society 25th Field Conference, 9:175-178.

—, 1989. The Triassic Dinosaur *Coelophysis*. Museum of Northern Arizona Bulletin 57:i-xv, 1-160.

—, 1990. Variation in *Coelophysis bauri*. in Carpenter, K. and Currie, P.J., eds, Dinosaur Systematics: Approaches and Perspectives, Cambridge University Press, Cambridge, p. 81-90.

Cope, E.D., 1889. On a New Genus of Triassic Dinosauria. American Naturalist, 23:626.

Padian, K., 1986. On the Type Material of *Coelophysis* Cope (Saurischia: Theropoda) and a New Specimen from the Petrified Forest of Arizona (Late Triassic: Chinle Formation). in Padian, K. (ed.), The Beginning of the Age of Dinosaurs, Cambridge University Press, Cambridge, p. 45-60.

Carnegie Museum of Natural History, *Coelophysis*, http://www.clpgh.org/cmnh/jurassic/fctcoelo.html

In Hand Museum, *Coelophysis bauri*, http://www.inhandmuseum.com/Dinosauria/Saurischia/Theropoda/Ceratosauria/Coelophysidae/Coelophysis.html

New Mexico Museum of Natural History, Dinosaurs in New Mexico, http://www.nmmnh-abq.mus.nm.us/nmmnh/dinosinnm.html

NEW YORK

EURYPTERUS REMIPES

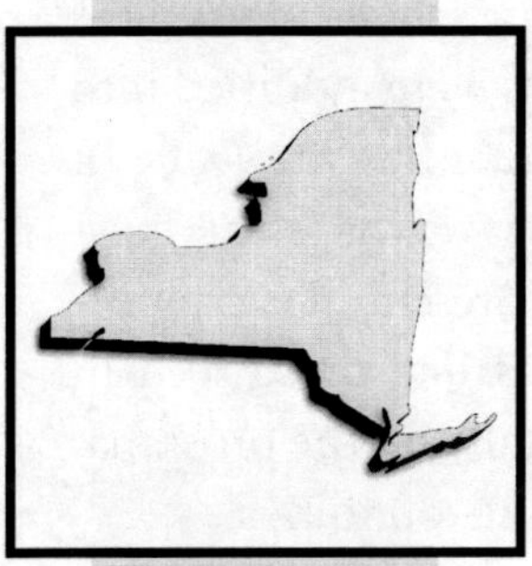

New York, known as the Empire State, has designated one of the most unusual animals to have ever lived as its state fossil—*Eurypterus remipes*, popularly called a sea scorpion.

Eurypterus remipes is known to science as a eurypterid, one of an extinct group of arthropods distantly related to the modern king crab, horseshoe crab, and land scorpion. *Eurypterus* inhabited the lands that make up New York some 410 million years ago, during the Silurian Period, when the state was covered by a vast sea. During this time the world was going through a period of immense change, with plants beginning to colonize the land and coral reefs spreading throughout the world's oceans. The New York area was no exception. A plethora of reefs sprang up in the vast sea that covered this area, and before long, *Eurypterus remipes* was the undisputed king of these reefs.

Exactly what made *Eurypterus* such a successful animal? Fortunately, this question can be answered by paleontologists. Because of the exquisite preservation in many sea scorpion fossils, more is known about the external morphology of eurypterids than perhaps any other group of fossil invertebrates. As in all arthropods, the body of *Eurypterus* was covered in a hard exoskeleton, which was furthermore segmented into two basic parts: the front prosoma and the rear opisthosoma.

Attached to the prosoma were a pair of compound eyes, the mouth, and six pairs of legs. The opisthosoma was further divided into a region of twelve tail segments and the telson, a triangular flap of armor extending from the tail.

In *Eurypterus*, there were six pairs of appendages attached to the prosoma. The first were the chelicerae, which are analogous to the pincers of scorpions. In most sea scorpions, including *Eurypterus*, the chelicerae were rather small. The next four pairs of appendages were the walking legs, which often were ornamented with hard spines. The final pair of appendages were modified into paddles, used for swimming. The mouth appeared on the soft ventral side of the prosoma, while the gills appeared on the opisthosoma. Perhaps the most unusual feature of *Eurypterus* was its eyes. Four eyes were present in each individual, two being simple eyes, and two being compound like those seen in many modern insects. Usually, the eyes were located centrally on the prosoma, which enabled the individual to have a clearer view of its surroundings.

The oldest eurypterid fossils come from the Lower Ordovician of New York, and are often very small. However, in a relatively short period of time these sea scorpions literally exploded in size, growing from an initial average of 1-2 feet long to an astonishing 7 feet in length by the Silurian, making them the largest arthropods ever known. Although the reason for the evolution of gigantism is unknown, it is thought by many that a larger body size made *Eurypterus*, and its fellow sea scorpions, more efficient predators. Before fishes and other

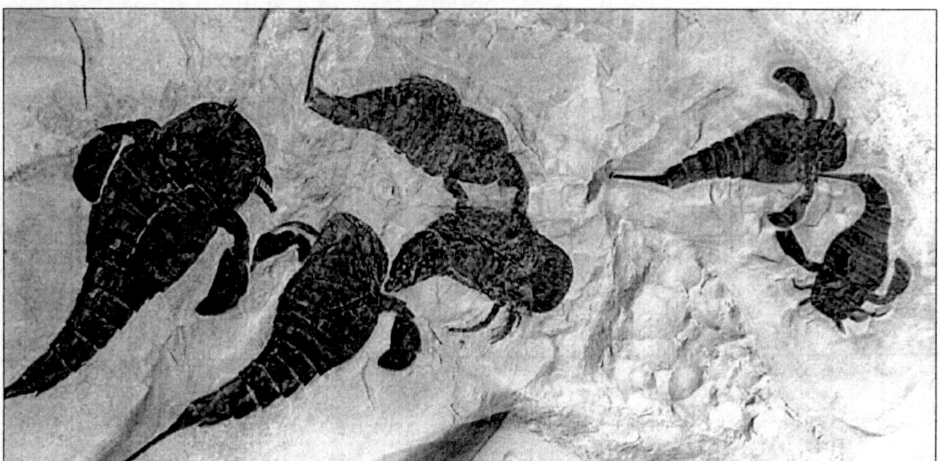

marine vertebrates perfected the art of hunting, sea scorpions terrorized the Silurian seas. Thought to be bottom dwellers, *Eurypterus* and its kin probably crawled along the ocean floor searching for food, which included trilobites, cephalopods, and whatever their pincers could grab.

Most eurypterid fossils, including those of *Eurypterus*, have been found in rock layers that were formed in brackish or freshwater environments, making the name "sea" scorpion a bit of a misnomer. However, many scientists do argue that primitive sea scorpions first evolved in the oceans, and only later moved into freshwater environments. It is even thought that some sea scorpions made their way into inland rivers, and possibly onto the land, before becoming extinct.

This extinction is a debate within itself. While eurypterids reached maximum diversity during the Silurian and Early Devonian, they disappeared about 245 million years ago, at the end of the Permian Period. At this time some 90% of marine life disappeared, possibly due to an asteroid impact, massive volcanism, or severe habitat disruption caused by the breakup of the Pangaean supercontinent. While the cause is debatable, it is likely that the disappearance of eurypterids is intertwined with the extinction of trilobites and other forms of marine life.

There is no debate, however, over the intrigue of sea scorpions. The first eurypterid fossil ever discovered was found in 1818 by fossil hunter S.L. Mitchell in Silurian rocks of New York state. He initially mistook the specimen for a catfish, and it was not until 1825 that eurypterids (meaning "broad wings") were recognized as a group of arthropods. Since Mitchell's initial discovery, thousands of eurypterid fossils have been discovered on every continent, and over 300 species have been named. These species have been classified into more than 60 genera and about 20 families.

Many of these thousands of fossils have been discovered in New York state. The majority of these fossils have been discovered in a Silurian rock unit known as the Bertie Dolostone. This dolomite layer extends from Buffalo to Schenectady and south to Poughkeepsie, roughly the route of the New York Thruway. Part of this unit is whimsically dubbed the "eurypterid beds" or the "sea scorpion graveyard." These beds begin just south of Utica, at the western end of the Mohawk Valley, and extend west to Buffalo.

Rocks composing the Bertie Dolostone can be seen as one travels the Thruway. One of the better spots to view them is an area of roadcuts located along the expressway on the west side of Rochester. Furthermore, a roadcut on the Lehigh Valley Railroad, less than a mile east of Cross Roads Station, passes through a bed of gypsum known as the Thompson Quarry. Above the gypsum in the banks east of the pit are about 25 feet of Salina Limestone, which contain an abundant array of fossils, including fragments of *Eurypterus*. Another well-documented eurypterid locality is Alan Lang's Quarry in Herkimer County. This Late Silurian quarry has yielded several fine sea scorpion specimens, including those of *Eurypterus remipes.*

However, complete eurypterid specimens are rare, and *Eurypterus remipes* is not as common as other New York eurypterid fossils. Regardless, it was named the official state fossil because it was the first sea scorpion ever discovered. Former New York State Paleontologist Donald Fisher was instrumental in campaigning for the adoption of *Eurypterus remipes* as the state fossil, and in 1984 a bill was signed by Governor Mario Cuomo making his dream a reality.

While well-preserved eurypterids are rare, several museums boast fine sea scorpion specimens. Among these museums are the American Museum of Natural History in New York City, the Museum of Science in Buffalo, the New York State Museum in Albany, the Heroy Geology Laboratory of Syracuse University, the Paleontology Research Institution in Ithaca, the National Museum of Natural History in Washington, D.C., the Denver Museum of Nature and Science, the Burpee Museum of Natural History in Rockford, Illinois, and the Pratt Museum of Geology in Amherst, Massachusetts.

Whether or not New York is truly a strange state, as some assert, is debatable. However, the rarity and uniqueness of its state fossil, *Eurypterus remipes*, is never argued. This Silurian sea scorpion, although rare, is well-known and well understood. And, fortunately for collectors, good specimens still remain in the Bertie Dolostone and other New York rock layers.

FOR FURTHER INFORMATION:

Brusatte, S., 2000. Kids' Corner: *Eurypterus remipes.* Fossil News: Journal of Avocational Paleontology, vol. 6, no. 9, p. 10.

Ciurca, S.J., Prehistoric Pittsford: The Silurian Eurypterid Fauna, http://www.geocities.com/paleoresearch/

Kjellesvig-Waering, E.N., 1979. Eurypterida. in: the Encyclopedia of Paleontology pp. 290-296 (Fairbridge and Jablonski, eds.), Dowden, Hutchinson, and Ross, New York.

Stormer, L. 1955. Merostomata. pp. P4-P41 in Moore, R.C. (ed.) Treatise on Invertebrate Paleontology. Part P: Arthropoda 2: Chelicerata. Geological Society of America and University of Kansas Press, Lawrence, Kansas.

University of California, Berkeley, Eurypterida, http://www.ucmp.berkeley.edu/arthropoda/chelicerata/eurypterida.html

NORTH DAKOTA

TEREDO PETRIFIED WOOD

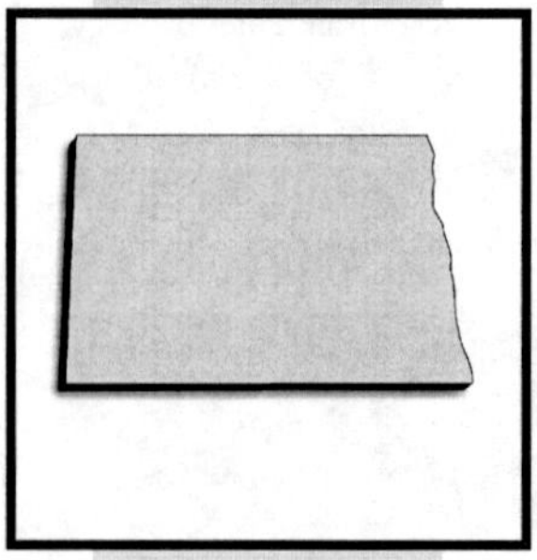

North Dakota is well-known for its dinosaur remains, which include many Cretaceous species. However, the state fossil of North Dakota is not a dinosaur, but rather a very peculiar type of Paleocene petrified wood. This petrified wood, called Teredo petrified wood, contains evidence of prehistoric shipworms, the same creatures that spell havoc for shipbuilders today.

During the Paleocene Epoch, about 60 million years ago, the terrestrial fauna of North Dakota was much different than that dominated by the dinosaurs some five million years earlier. Instead of being centered around large *Tyrannosaurus* and *Triceratops* individuals, this fauna was dominated by 50-foot-long crocodiles and giant, six-foot-tall birds known as *Diatryma*. North Dakota also was home to a vast marine fauna, as the state was mostly a lowland containing huge swamps and forests during much of the Paleocene. This environment, which had a subtropical to tropical climate, was not too different from areas of south Florida today.

On the edges of these swamps and lowlands was a large sea, today called by geologists the Cannonball Sea. This seaway, which was part of a vast, shallow epicontinental system, was the last ocean to cover North Dakota. A plethora of fossils is found in rocks deposited in the Cannonball Sea, including those of bony fish and sharks, crabs, lobsters, snails, and clams. Also found in

Cannonball Sea sediments are strange pieces of petrified wood, which contain evidence of prehistoric parasites.

The shipworm-infested wood found in North Dakota does not all belong to the same species. Rather, this wood belongs to a variety of bald cypress, ginkgo, redwood, and magnolia trees, all of which grew in the Everglade-like swamps and lowlands that bordered the Cannonball Sea. While these woods are not closely related in a phylogenetic sense, they are all coveted by collectors and scientists because of the markings they bear: those of shipworms.

The name "shipworm" is actually a misnomer. In reality, shipworms are not worms at all, but worm-shaped molluscs, related to clams, mussels, and oysters. Like these other bivalves, shipworms possess two shells. These shells are covered by ridges, which rasp wood as the shell rocks back and forth. This back-and-forth motion is powered by alternating contractions of the strong anterior and posterior adductor muscles. Between both shells is a pseudo foot, which acts as a suction disc. As the animal rasps, the disc is adjusted to slowly rotate the shipworm's body, resulting in a circular burrow. After this burrow is completed, the shipworm lines its home with a substance that hardens into a protective, limey tube. This hole is then sealed by two hard, paddle-like structures called pallets, which are located at the rear of the shipworm's body. Two tubes, or siphons, between the pallets draw in food and expel wastes. Like other molluscs, the shipworm is a filter feeder, and picks its food from water that is drawn in by the siphons.

Shipworms still exist in today's world, and regularly cause damage to

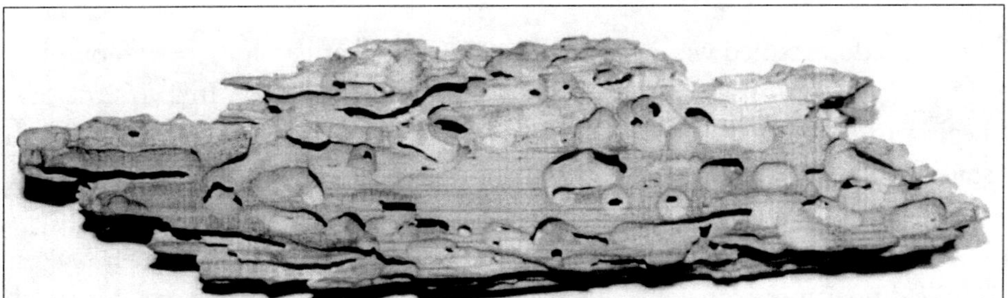

wooden ships. Scientists classify these molluscs in the Family Teredinidae, and give the animal the genus name *Teredo*, hence "Teredo petrified wood."

However, *Teredo* shipworms are not the only molluscs that can bore holes in wood. Several other mollusc species use their strong adductor muscles and powerful shell valves to burrow through wood and digest cellulose. Because of this, the name "Teredo petrified wood" is very confusing. While the majority of the wood discovered with bored holes is likely due to a *Teredo* shipworm, one can never be sure. In this sense, Teredo petrified wood is also a bit of a misnomer.

Also confusing is that Teredo petrified wood is not always petrified. In some cases, the wood has been replaced by mudstone or sandstone, leaving a cast full of holes. Often, these holes can be ascribed to burrowing molluscs, but sometimes their true nature remains unclear. Many a time has a fossil hunter uncovered a hole-filled rock thinking it was Teredo petrified wood, only to discover later that the specimen was only a simple piece of sandstone. Because of this confusion, the names "Teredo-bored petrified wood," "shipworm-bored fossil wood," and even "clam-bored fossil wood" have been suggested for the North Dakota state fossil.

Teredo petrified wood is one of the most common and well-known North Dakota fossils, and is known almost exclusively from the Cannonball Formation, a rock unit named after the prehistoric Cannonball Sea. Although this unit is quite large, the best Teredo petrified wood sites are located in the south-central portion of the state. The most productive localities are near Bismarck and Mandan, while good sites are also located southwest of these cities.

Teredo petrified wood became the state fossil after a long period of lobbying by members of North Dakota mineral and gem clubs. In 1967 the State Legislature rewarded the efforts of these clubs by making the unique wood the state fossil, becoming one of the first states to designate such a fossil.

Teredo petrified wood is a perfect symbol for the state of North Dakota. While dinosaur remains receive more publicity, Teredo petrified wood is sought out by collectors and rockhounds, who often polish the wood and make it into bookends and paperweights. And, in addition to being popular, Teredo petrified wood is also unique. This strange wood is bored by animals that still live, and wreak havoc, today. The bane of mariners of past centuries—and modern shippers with wooden boats—this small, worm-like mollusc has created a joy for fossil hunters.

FOR FURTHER INFORMATION:

Bartsch, P., 1921. A New Classification of the Shipworms and Descriptions of Some New Wood Boring Mollusks. Biological Society of Washington, Proceedings 34(1): 25-32.

—, 1922. A Monograph of the American Shipworms. United States National Museum Bulletin 122, 51 p., 37 pls.

Turner, R.D., 1966. A Survey and Illustrated Catalogue of the Teredinidae. The Museum of Comparative Zoology, Harvard University, Cambridge, Massachusetts, 265 p.

OHIO

ISOTELUS

Although many may not know it, Ohio has long been renowned for its invertebrate and vertebrate fossil record. While no dinosaurs have been discovered in the Buckeye State, countless marine invertebrates have. Among these invertebrates are crinoids, brachiopods, gastropods, and trilobites. One of these trilobites, *Isotelus*, is among the largest trilobites known. Because of its uniqueness, *Isotelus* has been officially designated as Ohio's state fossil.

Isotelus lived during the Ordovician Period, some 500-420 million years ago. During this time, Ohio was situated approximately 20° south of the equator, about where northern Australia is today. Warm, shallow seas covered much of the state, teeming with invertebrates and early vertebrates. The king of these seas was *Isotelus*.

Like other trilobites, *Isotelus* had a body section divided axially into three parts—the cephalon, thorax, and pygidium—and composed of a medial central axis and two lateral lobes ("trilobite" = "three-lobed"). Trilobites were arthropods, and hence were covered with an exoskeleton and possessed jointed legs. To grow, trilobites had to periodically molt; one trilobite may have left many shed carapaces to fossilize, explaining the diverse trilobite fossil record.

However, *Isotelus* differed from other trilobites in many ways. First, it was rather plain in appearance, and its cephalon and pygidium were similar in size and shape. Its exoskeleton was generally smooth, and it lacked the exorbitant ornamentation that characterized some other trilobites. However, what *Isotelus* lacked in style it made up for in size. While the average trilobite is only one or two inches long, *Isotelus* grew to lengths of 28 inches.

Not only was *Isotelus* big, but it was also carnivorous. It probably crawled along the sea floor, burrowing into the sediment and periodically emerging to hunt. Before the radiation of vertebrates, *Isotelus* was one of the most successful marine predators. As one scientist put it, *Isotelus* was the *Tyrannosaurus rex* of the Ordovician seas!

As a group, trilobites first evolved about 570 million years ago, during the Cambrian Period. *Isotelus* lived in Ohio about 445 million years ago, when the trilobites were attaining maximum diversity. However, these marine arthropods died out somewhat suddenly about 245 million years ago, at the end of the Permian Period. While their disappearance is somewhat of a mystery, many believe that either a large environmental disruption or the evolution of successful predators (such as horseshoe crabs) rang the death knell for trilobites. By any measure, however, 325 million years is successful.

Today trilobites are known only as fossils, with about 20,000 species currently recognized by paleontologists. Of all these species, *Isotelus* remains one of the largest.

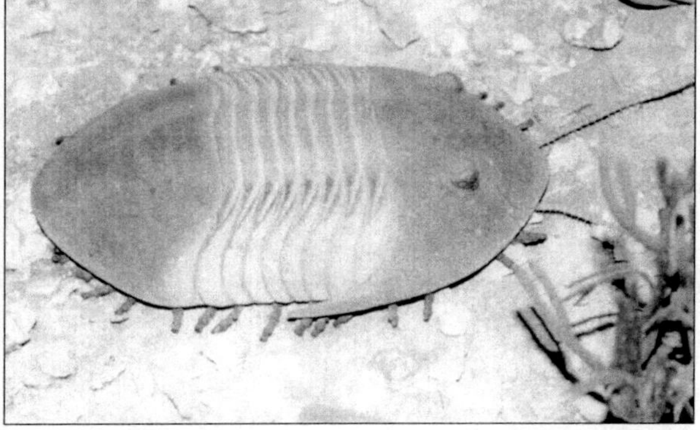

The history of this predatory giant in Ohio is almost as intriguing as the evolution and extinction of trilobites as a whole. The first serious study of Ohio's Ordovician rocks was undertaken by the Geological Survey of Ohio in 1837-1838. At this time, state geologist John Locke mapped a portion of the southwestern corner of the state. Among his many discoveries were partial remains of a large specimen of *Isotelus*. Locke collected only the pygidium of the trilobite, but he estimated that the entire specimen would have been about 21 inches in length. Because of its large size, Locke named the specimen *Isotelus maximus.* He later changed the name to *Isotelus megistos,* but today *I. maximus* is the accepted species name.

However, Locke's discovery pales in comparison with another *Isotelus* specimen. While digging an outlet tunnel during the construction of the Huffman Dam on the Mad River northeast of Dayton, workmen discovered a complete specimen of *Isotelus* in 1919. This specimen measured 14.5 inches long, and was 10.25 inches in width. After seeing the specimen, Arthur E. Morgan, the chief engineer of the Miami Conservation District, gave it to Dr. August F. Foerste, a paleontologist and physics teacher at Steele High School in Dayton. Realizing that it differed from other Ohio trilobites, he named it *Isotelus brachycephalus*. Foerste, who was a research associate at the National Museum of Natural History, donated the specimen to the Smithsonian, where it still occupies a permanent position in the National Museum fossil hall.

The Smithsonian displayed the specimen as the largest complete trilobite ever discovered, but had to change its description after another *Isotelus* specimen was discovered in Ohio. Found by professional trilobite hunter Tom Johnson, this complete specimen measures 16 inches long, and was discovered in Montgomery County in east-central Ohio.

Because of these finds, well documented in the news, several people pushed for *Isotelus*' designation as Ohio's official state fossil. Foremost among them were Doris Swabb's third grade students at Beavertown School in Kettering, and Virginia Evers' fourth grade students at St. Anthony School in Dayton. After visiting the Dayton Museum of Natural History and viewing a cast of the famous Huffman Dam specimen, the students and teachers came up with the idea of designating the Huffman Dam trilobite the official state fossil of Ohio. The students wrote to state representatives Robert L. Corbin and Robert E. Hickey of Dayton, who agreed to sponsor legislation in the Ohio House of Representatives to make *Isotelus* the official state fossil. State senator Charles Horn, also of Dayton, agreed to do the same in the Ohio Senate. The bill, which only specified the genus *Isotelus* as the official state invertebrate fossil, quickly passed both houses, and was signed into law by the Governor on June 20, 1985.

While complete *Isotelus* specimens are rare, discoveries of partial exoskeleton pieces are common in Ohio. These specimens are entombed in Ordovician rocks making up the Cincinnati Dome, an enormous area in southwestern Ohio where the earth has been arched and uplifted. These Ordovician rocks make up about 820 feet of comparatively thin, alternating layers of limestone and shale, which were deposited as limey mud and clay on the floor of the shallow seas. Many of the best sites are on private property, and the landowner's permission must be obtained before collecting.

However, good *Isotelus* specimens are exhibited in many museums, among them the National Museum of Natural History, the Dayton Museum of Natural History, the Cleveland Museum of Natural History, and the Orton Geological Museum of Ohio State University. There are also several publications available from the Ohio State Geological Survey which contain good descriptions, photos, and diagrams of *Isotelus*.

Although lacking in ornamentation, *Isotelus* remains as one of the largest trilobites discovered. And, although there are no large dinosaur fossils, the state of Ohio remains one of the most paleontologically diverse states in the country. It is fitting, then, that *Isotelus* is the official state fossil of the Buckeye State.

FOR FURTHER INFORMATION:

Behrendt, M., 2000. Feature fossil: *Isotelus latus.* Fossil News: Journal of Avocational Paleontology, v. 6, no. 5, p. 3.

Brusatte, S., 2001. Kids' Corner: *Isotelus.* Fossil News: Journal of Avocational Paleontology, v. 7, no. 5, p. 11.

Eckert, A. W., 1961. The mammoth trilobite of Dayton. Science Digest, July, p. 67-70.

Feldman, R.M. and Hackathorn, M., 1996. Fossils of Ohio. Ohio State Geological Survey, 577 p., 232 figs., 12 tables, glossary, index.

Foerste, A. F., 1919. Notes on *Isotelus, Acrolichas, Calymene*, and *Encrinus*. Denison University Scientific Laboratories Journal, v. 19, p. 65-82.

Fuchs, R., Some Trilobites of the Cincinnatian, http://www.geocities.com/drydredgers/trilobit.htm

Gnidovec, D.M., State fossil ruled ancient seas. Columbus Dispatch, March 23, 1997.

Hansen, M. C., 1985. *Isotelus*—Ohio's State Fossil. Division of Geological Survey, Ohio Geology, Summer, p. 1-4.

Locke, J., 1838. Geological report (on southwestern Ohio). Ohio Division of Geological Survey Second Annual Report, p. 201-286.

—, 1842. On a new species of trilobite of very large size *(Isotelus megistos)*. American Journal of Science, v. 42, no. 2, p. 366-368.

Shrake, D. L., 1994. Ohio trilobites. Ohio Division of Geological Survey GeoFacts No. 5.

Ohio State Geological Survey, *Isotelus*: Ohio's State Fossil, http://www.dnr.state.oh.us/odnr/geo_survey/geo_fact/geo_f06.htm

OKLAHOMA

SAUROPHAGANAX MAXIMUS

Oklahoma is an incredibly diverse state when it comes to paleontology. A well-known invertebrate fauna is found in a series of rock units that outcrop over much of the state, Permian reptiles and mammal-like reptiles are commonly discovered, and a plethora of dinosaurs and dinosaur footprints are found in the state's Morrison Formation sediments. However, it was not until 2000 that Oklahoma finally designated a state fossil. The logical choice was a very large, carnivorous dinosaur.

About 150 million years ago, during the Jurassic, Oklahoma was home to a diverse dinosaur fauna. A series of rivers, streams, and small seas ran through the landscape, assisting the preservation of dinosaur fossils. To this date, hundreds of dinosaur skeletons have been plucked from the plains of the Sooner State, including those of *Apatosaurus, Allosaurus*, and the state fossil, *Saurophaganax*.

Saying *Saurophaganax* was large is an understatement. This theropod grew to a height of 16 feet and a length of 40 feet, rivaling *Tyrannosaurus, Giganotosaurus*, and *Carcharodontosaurus*. However, unlike *Tyrannosaurus, Saurophaganax* was equipped with long arms and three-clawed hands, which allowed the dinosaur to grasp and hold its prey, including small sauropods and ornithopods. The claws of *Saurophaganax* were among the largest of any theropod.

Saurophaganax was a member of the allosaurid lineage, and many have suggested that this dinosaur may indeed be an overgrown *Allosaurus*. However, paleontologists have offered three distinct morphological characteristics that set *Saurophaganax* apart from *Allosaurus*. For starters, *Saurophaganax* possessed expanded chevrons on the tail vertebrae, which were more derived than those of *Allosaurus*. Secondly, *Saurophaganax* was signaled by the presence of lamina (a tissue membrane) along the base of its neural spine. Finally, *Saurophaganax* possessed an atlas that is very much unlike that of *Allosaurus*. But, this information is much of what we know about *Saurophaganax*. This dinosaur remains poorly understood, and its remains are rare and isolated.

Despite its rarity and the fact that science barely understands it, *Saurophaganax* became Oklahoma's state fossil on April 14, 2000, when Governor Frank Keating signed Senate Bill 1185. The bill was sponsored by State Senators Cal Hobson and Dave Herbert, and State Representative Tommy Thomas.

Strangely enough, this bill states, "This spectacular dinosaur, the greatest king of reptile eaters, once roamed this great land. It is only known from Oklahoma and has surpassed the *Tyrannosaurus rex,* the 'king of dinosaurs,' as the greatest predator of earth's history." While "the greatest king of the reptile

eaters" refers to the etymology of *Saurophaganax*' name, the *Tyrannosaurus* remark is most likely wrong. Paleontologist James Farlow addressed this fact in a 2000 Internet posting, in which he described his study of *Tyrannosaurus* and *Saurophaganax* feet. *Saurophaganax*' first phalanx of digit one measured 93 millimeters, Farlow said. This seems paltry in comparison to that of *Tyrannosaurus rex.* The smallest *T. rex* first phalanx measured 193 millimeters, while the largest measured 206 millimeters. To quote Dr. Farlow, "the numbers ain't even close." Regardless, many still do believe that *Saurophaganax* was larger than *Tyrannosaurus rex,* a notion which is most likely wrong.

Saurophaganax is quite rare, and its remains have been found only in the Morrison Formation of Oklahoma. More specifically, all known *Saurophaganax* bones come from a quarry in Cimarron County, near Black Mesa, located near the town of Kenton. The first known specimens of this dinosaur were discovered in the late 1930s by a WPA crew working under the direction of J. Willis Stovall, a University of Oklahoma paleontologist.

In 1941, the first description of these fossils was published by G.R. Ray in a *Natural History* article entitled "Big For His Day." In this article Ray referred to the dinosaur as *Saurophagus maximus*, but provided no detailed descriptions to accompany the illustrations included with the article. According to most published reports, it was Stovall who suggested this name to Ray, based on a specimen, likely a tibia described as a "hefty long shank," which was catalogued in the Oklahoma University State Museum as specimen number 4666.

However, the name *Saurophagus* turned out to be preoccupied, being originally ascribed to a different dinosaur in 1831. In addition, specimen number 4666 was apparently lost, as the Oklahoma Museum's vertebrate fossil catalogue does not have a record of the old specimen. This presented a very difficult nomenclatural situation. Some of it was clarified in a 1995 paper written by Dinosaur National Monument paleontologist Daniel Chure. In this paper, Chure chose a new genus for this dinosaur—*Saurophaganax*, which as previously stated, means "king of the reptile eaters." Chure chose to retain the species name, *maximus*, which translates as "greatest," hence, "greatest king of the reptile eaters." Chure also concluded that, based on the number of metatarsals and femora collected in the Cimarron County quarry, parts of at least two individuals of equal size were discovered, which is good news for scientists working on this very fragmentary dinosaur.

But, as it turned out, the nomenclatural situation remained tenaciously murky. Often, the original Stovall "Saurophagus" material is referred to the genus *Saurophaganax*, and it certainly does show some similarity to it. However, the original "Saurophagus" is probably intermediate, and thus cannot be reasonably made a senior synonym of any species, including *Saurophaganax*. In addition, Chure also judged the lectotype of "Saurophagus," published on by Hunt and Lucas in 1987, to be undiagnostic. Therefore, despite the similar species name and the superficially similar remains, *Saurophaganax* is *not* intended as a replacement name for "Saurophagus." The two dinosaurs may indeed be one and the same, but scientific nomenclature does not designate them as such, making "Saurophagus" at least technically valid. However, the Oklahoma State Legislature did the wise thing in designating *Saurophaganax* the official state fossil.

Despite all of this nomenclatural mess, visitors can still enjoy a wonderfully-produced cast skeleton of *Saurophaganax* at the new Sam Noble Oklahoma Museum of Natural History, on the campus of the University of Oklahoma in Norman. This new museum, open to the public on May 1, 2000, contains 195,000 square feet of exhibit space, making it the largest university-based natural history museum in the world. Much of this exhibit space is taken up by the Museum's Hall of Ancient Life. Included in this Hall is a "Clash of the Titans" exhibit, which depicts a *Saurophaganax* individual squaring off with an *Apatosaurus*. The *Saurophaganax* in this exhibit was cast by Dinolab in Salt Lake City, and is 39 feet long and 16 feet high. A similar *Saurophaganax* cast was included in the Discovering Dinosaurs exhibition in Wilmington, Delaware, while another was also showcased at Dinofest 2000 in Chicago. Visitors may also view a *Saurophaganax* cast at the Utah Museum of Natural History.

Despite the mess surrounding its discovery and name, *Saurophaganax* is a great state fossil for Oklahoma, a state as unique and diverse as any other. While little is known about this dinosaur, it certainly was huge—and a very important component of the ecosystem of Jurassic Oklahoma. We can all be glad that this beast has long since disappeared from the Sooner State's plains.

FOR FURTHER INFORMATION:

Brusatte, S., 2002. Kids' Corner-*Saurophaganax*. Fossil News: Journal of Avocational Paleontology, v. 8, no. 4, p.11.

Chure, D.J., 1995. A reassessment of the gigantic theropod *Saurophagus maximus* from the Morrison Formation (Upper Jurassic) of Oklahoma, USA. Sixth Symposium on Mesozoic Terrestrial Ecosystems and Biota, p. 103-106.

Coldwell, L., Legislature Names *Saurophaganax* at OU Museum State Fossil, Sam Noble Museum of Oklahoma History Website, http://www.snomnh.ou.edu/mediarelations/statefossil

Ray, G.R., 1941. Big For His Day. Natural History.

Williams, G.R., 2000. The Sooner State's New Showcase of Paleontology-A Report on the Sam Noble Oklahoma Museum of Natural History. Dinosaur World, no. 9, p. 5-11.

Sam Noble Oklahoma Museum of Natural History, http://www.snomnh.ou.edu/

PENNSYLVANIA

Phacops rana

Mesozoic rocks are commonly exposed in the western United States, with amazing assemblages of Triassic, Jurassic, and Cretaceous rocks known from such states as New Mexico, Utah, and Wyoming. Ordovician and Silurian rocks are common in Midwestern states like Ohio and Indiana, while Pennsylvanian sediments commonly outcrop in Illinois and Missouri. However, one geologic age isn't as well-represented as the above groups. Although the Devonian was a period of immense global change, Devonian rocks and fossils are quite rare in many parts of the United States. Fortunately, this is not the case in Pennsylvania, a state that is so rich in Devonian rocks that it has designated a Devonian specimen as its state fossil.

During the Devonian, some 405 to 365 million years ago, the first semi-modern ecosystems were beginning to evolve, and plants were getting their first foothold on land. Animals were also beginning to colonize the land, with amphibians and insects occupying most terrestrial niches. Strange animals were also continuing to evolve in the sea. Foremost among these animals were the trilobites, a group of incredibly diverse arthropods. One of the most common of these trilobites, *Phacops rana*, is designated the Pennsylvania state fossil.

Like other trilobites, *Phacops rana* is an arthropod, or an animal with jointed appendages and an exoskeleton. All trilobites possessed a body that consisted of few to many segments, most of which bore leg-like appendages. The exoskeleton was divided into three longitudinal lobes, a central lobe extending from head to tail and two less prominent pleural lobes which flanked the central lobe. Most trilobites, including *Phacops*, possessed exoskeletons divided into

three transverse sections—the cephalon, the head segment which usually bore eyes and antennae; the thorax, or flexible middle section; and the pygidium, or tail section, which was fused and rigid.

The mouth was located at the bottom rear of the head, and the stomach was situated in front of the mouth, in the upper part of the head. This arrangement contrasts greatly with other marine organisms. Because of this, the digestive tract was forced into a J-shaped loop inside the head, which connected the mouth and the stomach. While this loop isn't preserved in fossils *per se*, scientists have hypothesized its position by comparing the trilobite's exoskeleton with that of other arthropods.

Trilobites also possessed complex eyes, which consisted of hundreds of lenses. As a group, trilobites were one of the first known organisms to have possessed true eyes. These arthropods, including *Phacops*, also had a well-developed nervous system, which included large antennae. The exoskeleton was composed of chitin, and was commonly molted, which meant that one individual trilobite could leave tens of skeletons, all of which were eligible for fossilization.

Phacops rana shared the above traits with other trilobites, but also possessed its own set of unique characters. As a genus, *Phacops* was characterized by indentations on the glabella, a large, circular section on the cephalon. Also on the cephalon were two large, kidney-shaped eyes. The thorax usually possessed 11 distinct segments and a gradually tapering axis. The pygidium was small and usually curved. Generally, *Phacops* specimens ranged from about 1-6 inches long.

Phacops is unique in that it is one of only a handful of genera that is found with well-preserved eyes and recognizable appendages. While appendages were present on every trilobite, *Phacops* is one of only four genera in which they have been well-studied. *Phacops* is also commonly found preserved in a rolled-up state, much like how modern pill bugs look after they have been disturbed. Why did *Phacops* roll up into a ball? Many trilobites show bite marks and other damage from predators. It is likely that when *Phacops* was disturbed, it rolled up into a ball, a position which would have provided protection for the trilobite's soft, ventral (belly) surface.

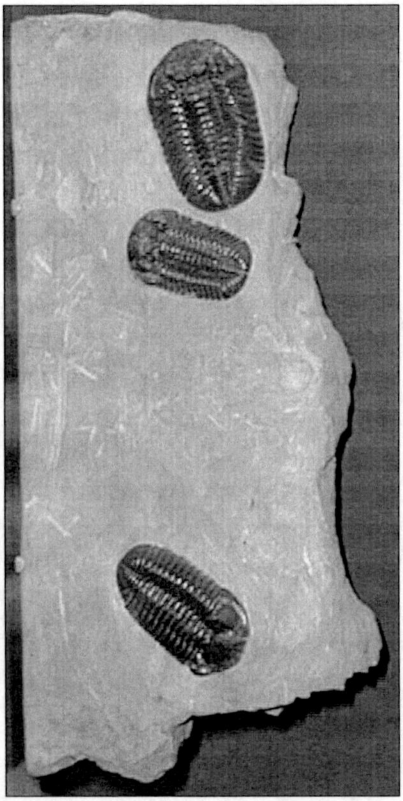

However, some scientists have pointed out that an enrolled trilobite would have also been easier to swallow than an extended one. Because of this, it has been suggested that, perhaps, a rolled-up trilobite was denser than an extended one, and rolling up would suddenly cause the individual to descend in the water column. This descent may have been enough to avoid predators. It is even possible that *Phacops* may have remained enrolled for quite some time. Features preserved on the cephalon and pygidium indicate that these two structures may have been able to remain linked for long periods. Perhaps, it has been suggested, *Phacops* may have been able to slow its body functions down during periods of food shortages by simply rolling up. This would have saved valuable energy.

Whatever the reason for this enrolling, the fact remains that numerous well-preserved *Phacops* specimens have been discovered in a rolled-up posture. Many of the best come from Pennsylvania, a state with a plethora of Devonian rock units. Many of the best *Phacops* specimens come from the Mahantango Formation. This formation regularly outcrops across central Pennsylvania, and is dated to the Middle Devonian. In fact, nearly 40% of the fossil sites listed for central and eastern Pennsylvania belong to the Mahantango Formation.

One of the better Mahantango sites is located in Lewis Township. It was inadvertently discovered by William E. Kochanov, a geologist, as he was searching for trilobite fossils in the Turbotville area of Northumberland County. Although he expected nothing to be found, Kochanov stumbled upon a Mahantango shale pit containing an abundance of well-preserved crinoids and trilobites, including *Phacops rana.* The site is owned by a Mr. Swope of Turbotville, so permission must be granted before collectors may enter.

Because of the abundance of specimens discovered, *Phacops rana* was designated the Pennsylvania state fossil by the General Assembly on December 5, 1988. Much of the credit for this designation goes to students from Valley Elementary School in Bensalem, Pennsylvania. However, when the students first began to think about a state fossil, they wanted a dinosaur. Upon learning that dinosaurs were only known from Triassic footprints in eastern Pennsylvania, they instead turned their attention toward *Phacops*, a fossil which reminded many of the students of Darth Vader! Later, the General Assembly honored their request by making this ubiquitous and unique trilobite the state fossil.

Little did the students know, however, that 13 years after their aborted quest to make a dinosaur the state fossil, the General Assembly would introduce a bill to designate an official state dinosaur. Known as *Atreipus milfordensis*, this dog-sized herbivore lived about 200 million years ago during the Late Triassic. Peter Kranz, a paleontologist instrumental in designating a Maryland state dinosaur, believes that it travelled in herds across the eastern United States. A bill to designate *Atreipus* as state dinosaur was introduced to the General Assembly in May, 2001.

Soon, Pennsylvania may have both a state fossil and a state dinosaur. However, it will always be the state fossil that garners more attention from collectors. Although Devonian sediments are somewhat rare in the United States, Pennsylvania is a lodestone for those studying this period of geologic time. This historic state is also a haven for those interested in Devonian fossils, including one of the most interesting and unique trilobites, *Phacops rana*, the official Pennsylvania state fossil.

FOR FURTHER INFORMATION:

Behrendt, M., 2001. Feature Fossil-*Phacops*. Fossil News: Journal of Avocational Paleontology, vol. 7, no. 2, p. 3.

Brusatte, S., 2001. Kids' Corner-*Phacops rana.* Fossil News: Journal of Avocational Paleontology, vol. 7, no. 9, p. 20.

Kochanov, W.E., and Inners, J.D., 1993. Mahantango Fossils. Pennsylvania Geology, vol. 24, no. 1. (viewed online at http://members.aol.com/jmfabiny/Mahantango.html)

Pellant, C., 1996. An Illustrated Guide to Fossils. The Nature Company, Surrey, Great Britain.

Struve, W., 1990. Palaozoologie III (1986-1990). Forsch Institute, Senckenberg, 127: 251-279.

Pennsylvania Geological Survey, *Phacops rana,* http://www.dcnr.state.pa.us/topogeo/phac.htm

The Mahantango "garden spot" is located in Lewis Township about 2 miles north of Turbotville. To reach the site, take the Turbotville exit off Interstate Route 180 and travel on Pa. Route 54 east towards Turbotville. Drive approximately 1 mile, turn north on SR(State Route) 1011 (Warrior Run Road), proceed about 1.4 miles, and turn east on SR 1010. The borrow pit lies on the north side of the road, approximately 0.2 miles from this last intersection. There is room to park one or two vehicles on the floor of the pit.

SOUTH DAKOTA

TRICERATOPS

South Dakota, like many other western states, is rich in dinosaur fossils. Bones, footprints, teeth, and other fossils from countless species of dinosaurs are found in the state, with one of the most famous being the *Tyrannosaurus rex* nicknamed Sue. However, *Tyrannosaurus* is not the official state fossil of South Dakota—or any other state, for that matter. Instead, the South Dakota State Legislature decided in 1988 to designate an herbivore that coexisted with this infamous predator: *Triceratops*.

That is not to say that *Triceratops* is not important. In fact, this dinosaur has given new insights into the evolution and extinction of dinosaurs, and may well have been one of the last dinosaur species to evolve. *Triceratops* is known from the Late Cretaceous, and lived about 68-65 million years ago, just before the dinosaurs became extinct. It is also one of the most recognizable dinosaurs, and is known by nearly every schoolchild as having a "three-horned face," the basis for its scientific name.

Triceratops was a member of the ceratopsian lineage, a group of dinosaurs which possessed frills and horns on their skulls. Of these ceratopsians, *Triceratops* is the largest, most common, and best-known, and although a complete skeleton has never been found, this dinosaur is known from several good skulls. The average *Triceratops* individual grew to lengths of at least eight meters (25 feet), and may have weighed upwards of 4.5 metric tonnes (5 English tons).

The most distinctive part of *Triceratops*' anatomy was undoubtedly its skull. This skull, which grew to lengths of eight feet and remains one of the largest skulls known from any land-dwelling animal, was characterized by three horns. Two of these horns were positioned behind the eyes, and are known as brow horns. The third horn, the nasal horn, was located above the nose. While the nasal horn was short, the brow horns often grew as long as three feet. All three horns were anchored by a bony horn core, an outgrowth of the skull bones, and covered in a tough mesh of keratin, the same material that makes up our fingernails and the horns of bulls. Also located at the rear of the skull was a large frill, composed of the parietal and squamosal bones, which grew to a length of six or seven feet. This is one of the only ceratopsian frills to lack parietal fenestrae (skull holes).

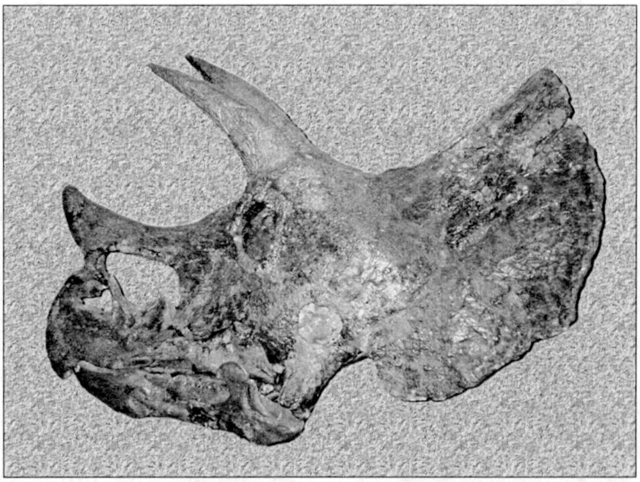

The jaws of *Triceratops* occupied the anterior portion of the skull. These jaws were strong, and equipped with teeth that looked much like elongated blades. Unlike human teeth, the teeth of *Triceratops* lacked the ability to grind or crush the plant material that the dinosaur fed on, but instead were used as shears. These scissor-like teeth were powered by strong jaw muscles, which were anchored to the large frill. Because of the uniqueness of this mastication system, it is thought that *Triceratops* did not eat ordinary leafy plant material, but instead tougher and more fibrous plants such as cycads or palm fronds.

The body of *Triceratops* was sturdy and strong, and was carried on four legs. There is debate over the proper orientation of these legs, with some scientists believing that they were carried straight up, while others point to a more sprawled stance. Regardless, *Triceratops* was likely rather slow at walking, but could charge like a rhinoceros. This charging may have been used to ward off predators, such as its contemporary, *Tyrannosaurus*.

The strange skull of *Triceratops* was probably also used for defense. Since the first skull was discovered in the 1880s, it was thought that *Triceratops* used its horns to defend itself against predators. One skull, discovered by paleontologist Chris Morrow, shows probable *Tyrannosaurus* bite marks, while another shows a broken brow horn with signs of rehealing. However, many paleontologists also believe that the skull of *Triceratops*, and other ceratopsians, was not primarily a defensive weapon, but rather a display structure used to attract mates. The skull may also have been used for intraspecific combat, possibly fighting for mates or territory, a notion which is supported by the discovery of puncture wounds on a set of skulls. A recent theory states that possibly the nasal horn was used for defense, while the brow horns were used for display. The real answer will likely never be known.

Some scientists have suggested that the horns of *Triceratops* may give clues to the gender of the animal. Since the first *Triceratops* specimen was discovered, sixteen species have been attributed to this genus. Based on observations of modern animals, it is unlikely that sixteen species of the same animal would inhabit the same area in the same general timeframe, a notion that is implied if indeed there truly were sixteen different species. In a 1986 paper, John Ostrom and Peter Wellenhofer rejected the notion of sixteen species, but instead recognized only one: *Triceratops horridus*, the first named *Triceratops* species. This study was supported by a 1990 paper written by Thomas Lehman. However, a different 1990 paper, authored by Cathy Forster, recognized two *Triceratops* species: *T. horridus* and *T. prorsus*. But, it is possible that Forster's two species represent males and females, or vice versa.

While most important *Triceratops* discoveries have actually taken place in Wyoming, one of the first records of this dinosaur can be traced to Colorado. In 1887, a pair of large fossil horn cores was sent to O.C. Marsh by Whitman Cross, a geologist who was studying the age and fossils of a rock unit near Denver. Marsh, going against the advice of some of his colleagues and collectors, concluded that the horn cores represented an extinct buffalo. In

December 1888, he officially described them as *Bison alticornis*. Later in 1888, collector John Bell Hatcher came across a large horn core while prospecting near Lusk, Wyoming. He wrote to Marsh describing the discovery, and later sent the professor the fossils. These horn cores finally convinced Marsh that what he called *Bison alticornis* was not a buffalo, but an extinct dinosaur. In 1889 Marsh officially named *Triceratops*.

More importantly, perhaps, this discovery launched the career of Hatcher as a horned-dinosaur expert. For the next few years, from 1889-1892, Hatcher spent much of his time in Wyoming, uncovering hordes of unknown ceratopsian skulls, among them more species of *Triceratops*. One of these species, *Triceratops prorsus*, was named in 1890. Many websites list *T. prorsus* as the official South Dakota state fossil, although the State Legislature only designated "*Triceratops*." Later, in 1927, a beautifully-preserved *Triceratops* specimen was discovered in Harding County, South Dakota. Today, it resides at the Museum of Geology in Rapid City. Specimens of *Triceratops* continue to turn up in South Dakota today.

Up until 1988, the official state fossil of South Dakota was the cycad. Cycads, gymnospermous plants often confused with ferns and palms, are marked by a central trunk topped by a whorl of leaves. These plants were prevalent in the Cretaceous, and were likely a food source for *Triceratops*. The most famous cycad fossil locality in the United States is near Minnekahta, in the Black Hills of South Dakota. Here, an entire forest of "cycads"—actually cycadeoids or bennettitaleans—was fossilized, and many specimens were later excavated and sent to museums around the world. At one time, the area was designated Fossil Cycad National Monument. However, vandalism and illegal collecting have denuded the area so badly that it has been closed as a national monument. These ravages also led the cycad to lose its state fossil designation.

While there have been few extremely important *Triceratops* discoveries in South Dakota, this dinosaur is still a fine state fossil. Well-known from Hell Creek sediments, this Late Cretaceous ceratopsian may have been one of the last dinosaurs to have evolved, and likely lived up to the K/T boundary. Today, only fossils are left, many of which are still discovered in South Dakota.

FOR FURTHER INFORMATION:

Bigelow, P., Hell Creek Faunal Facies, http://www.scn.org/~bh162/hellcreek2.html

Clos, L.M., 1996. Cycads: Living Fossils. Fossil News: Journal of Amateur Paleontology, v. 2, no. 11 (may be viewed online at http://www.fossilnews.com/1996/livingfossils.html).

Dodson, P., 1996. The Horned Dinosaurs. Princeton University Press, Princeton, New Jersey, 346 p.

Forster, C.A., 1990. The Cranial Morphology and Systematics of *Triceratops* with a Preliminary Analysis of Ceratopsian Phylogeny. Pennsylvania: University of Pennsylvania Department of Geology, unpublished Ph.D. thesis, 227 p.

Hatcher, J.B., Marsh, O.C., and Lull, R.S., 1907. The Ceratopsia. United States Geological Survey Monograph 49: 1-300.

Lehman, T.M., 1990. The Ceratopsian Subfamily Chasmosaurinae: Sexual Dimorphism and Systematics. in: Dinosaur Systematics-Approaches and Perspectives, Carpenter, K., and Currie, P.J., eds, Cambridge University Press, Cambridge, pp. 211-229.

Marsh, O.C., 1888. A New Family of Horned Dinosaurs from the Cretaceous. American Journal of Science, ser. 3, 36: 477-478.

—, 1889a. Notice of New American Dinosauria. American Journal of Science, ser. 3, 37: 331-336.

—, 1889b. Notice of Gigantic Horned Dinosauria from the Cretaceous. American Journal of Science, ser. 3, 38: 173-175.

Ostrom, J.H., and Wellenhofer, P., 1986. The Munich Specimen of *Triceratops* with a Revision of the Genus. Zitteliana 14: 11-158.

—, 1990. *Triceratops*: An Example of Flawed Systematics. in: Dinosaur Systematics-Approaches and Perspectives, Carpenter, K., and Currie, P.J., eds, Cambridge University Press, Cambridge, pp. 245-254.

TENNESSEE

PTEROTRIGONIA THORACICA

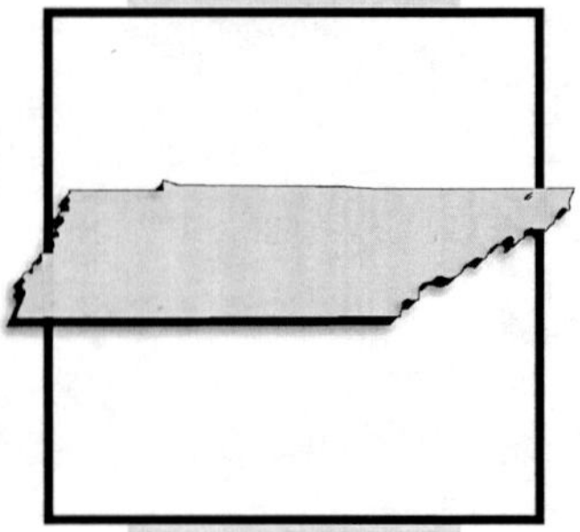

One of the most-visited geological wonders in the world calls Tennessee home. Of all of America's National Parks, the Great Smoky Mountain National Park garners the most visitors. This park, although not home to fossils *per se*, can boast some of the most interesting—and geologically historical—mountains in the Western Hemisphere. Now, after an intense period of lobbying, Tennessee can boast a new geological symbol: a state fossil. However, unlike the vast and omnipresent Smoky Mountains, this state fossil is nothing more than a small bivalve.

This bivalve, known to science as *Pterotrigonia (Scabrotrigonia) thoracica,* lived in Tennessee during the Cretaceous Period. During this time, much of the state was covered by a vast sea, which was home to a variety of marine life. Many of these were molluscs, and of these, *Pterotrigonia* was one of the most common.

Pterotrigonia, as mentioned above, is a bivalved mollusc, the group that contains clams, mussels, oysters, and scallops. Each one of these bivalves possesses a flattened body enclosed by two shells, or valves, hinged in the back, as well as a strong muscular foot used for digging burrows. The head of the bivalve is either reduced or missing completely. Also, most bivalves have large muscles used to close their shells. These muscles are the "meat" of the oysters and mussels commonly eaten by humans.

Bivalves are mostly sedentary filter-feeders and are usually found in marine environments. As a general rule, most bivalves prefer shallow water and tend to cluster in intertidal zones. It is thought that *Pterotrigonia* burrowed in shallow waters and was a suspension feeder like most of its close relatives.

When *Pterotrigonia* was alive, some 70 million years ago, seas covered most of western Tennessee. The aging Appalachian Mountains towered to the east of these seas, providing them with ample runoff water. This ever-replenishing supply of water provided the perfect environment for a slew of marine invertebrates, including snails, cephalopods, annelids, sponges, corals, crustaceans, and bivalves. Although many bivalves are found in the sediments that make up the remnants of these seas, *Pterotrigonia* is distinguished by its wedge-shaped shell and prominent ribs.

The vast beds of *Pterotrigonia* fossils are a testament to this animal's success. However, *Pterotrigonia* seems to disappear after the Cretaceous. Was the demise of this bivalve related to the great K/T extinction, the same event that wiped out the dinosaurs? This is very possible, and accepted by many scientists. Regardless, *Pterotrigonia*'s bivalve relatives survive today. Its closest relative, a bivalve named *Neotrigonia*, lives off the coast of New Zealand.

Specimens of *Pterotrigonia* are abundant in the Cretaceous rocks of west Tennessee, and are often found in vast beds, possibly signifying massive die-offs. These specimens are easily recognized, and because of their excellent state of preservation, are exhibited in many museums across the state. One museum in particular, the Coon Creek Science Center, is located near the spot where some of the best *Pterotrigonia* specimens have been unearthed. These specimens were discovered in the rock unit that gave the Museum its name: the Coon Creek Formation.

It is obvious that specimens of *Pterotrigonia* are abundant and popular with Tennessee residents. Because of this, Michael Gibson began lobbying the State Legislature to name this bivalve the state fossil in the late 1990s. Gibson, a paleontology professor at the University of Tennessee at Martin, urged his students to aid him in his quest. Over time, Gibson's drive became a project sponsored by the University's GeoClub, and was eventually supported by the Tennessee Earth Science Teachers Organization and numerous other groups.

Gibson recalled his reasons for wanting a state fossil in an online interview. His major objective, actually an ulterior motive, lay in education. Gibson argued that, "the K-12 framework for the State of Tennessee has standards and benchmarks directly related to fossils and earth history." However, despite the educational standards, Tennessee was in the minority of states in not having a state fossil. Gibson made it his goal to change this. He also listed other reasons for his campaign, including the fact that a state fossil would "increase public awareness of paleontology, represent and promote the fossil riches of Tennessee as a natural resource, become a tourist attraction, and represent the public understanding of the history of life on earth."

From 1996-1998 the GeoClub promoted these ideals, and in 1998 their dream became a reality. In May of that year, Governor Sundquist signed House Joint Resolution 552, officially designating *Pterotrigonia* as the state fossil. This bill, pushed for so energetically by the GeoClub, was sponsored by Representative Mark Maddox. This made Tennessee the 38th state to designate an official fossil. One year later, in 1999, Gibson and Don Byerly, a professor at the University of Tennessee at Knoxville, created a series of field-based workshops for K-12 teachers, using the new state fossil as a classroom tool. Gibson is also in the process of writing a book on *Pterotrigonia*.

While *Pterotrigonia* may only be a small, simple bivalved mollusc, this animal plays a very important part in the education of Tennessee students. Because of the hard work and tireless efforts of Gibson and his students, Tennessee can now boast a state fossil of its own, in addition to its geological jewel, the Great Smoky Mountains.

FOR FURTHER INFORMATION:

University of Tennessee at Martin, Tennessee State Fossil,
http://www.utm.edu/departments/artsci/ggp/statefossil/fossil.htm

UTAH

ALLOSAURUS

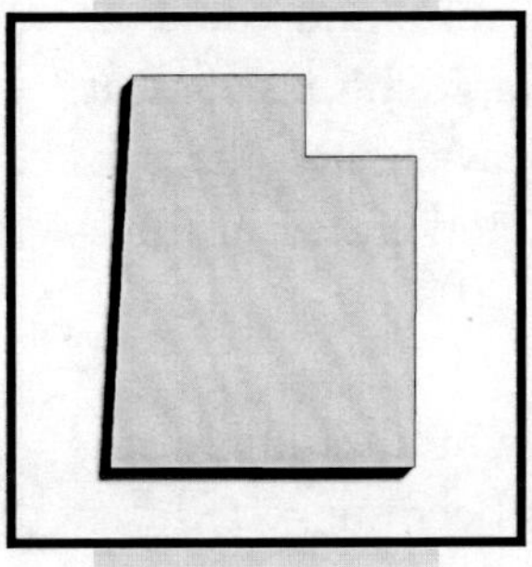

Dinosaur fossils have been discovered throughout the American west, from Montana and Colorado to Wyoming and Oklahoma. However, perhaps the state with the most diverse dinosaur fossil record is Utah. Rock units from all three periods of the Mesozoic are exposed in the Beehive State, and remains of several Triassic, Jurassic, and Cretaceous dinosaurs have been discovered. Despite this diversity, one Utah dinosaur reigns over all the rest. Because of its abundance and popularity, *Allosaurus* is designated as Utah's official state fossil.

Allosaurus is a large theropod known from the Late Jurassic, about 156-145 million years ago. During this time, much of Utah was a giant floodplain, full of lush vegetation and countless species of dinosaurs. While the state was covered by deserts earlier in the Jurassic, Late Jurassic Utah marked a profound shift in climate, as the area gradually began to cool off and become more hospitable. One of the most common dinosaurs that inhabited this land was *Allosaurus*. Today, *Allosaurus* remains are common in the Morrison Formation sediments that underlie and crop out over much of Utah.

Allosaurus is the best-known large carnivore of the Late Jurassic Period, and this ubiquity has given paleontologists new insights into the anatomy and predatory habits of this dinosaur. Hundreds of specimens have been discovered. The average *Allosaurus* individual measured somewhere from 9-12 meters (30-40 feet) long, and weighed somewhere near 1400 kilograms. Like all theropods, *Allosaurus* walked on two strong legs, which were large and powerfully muscled, and likely could have enabled the individual to run upwards of 12-20 miles an hour. *Allosaurus*' neck was also strong, and marked by an S-shape composed of vertebrae different from those of other dinosaurs, hence the name *Allosaurus* ("different lizard"). This neck tapered into a meter-long skull armed with more than 70 eight-centimeter-long teeth. The lower jaws of *Allosaurus* were hinged in front, and could expand slightly sideways, enabling the individual to swallow large chunks of meat. *Allosaurus* also possessed a massive tail, bulky body, and heavy bones. Its arms were short, although longer than those of *Tyrannosaurus*, and ended in three-fingered hands with six-inch-long claws.

The skull of *Allosaurus* was adorned with two short brow-horns (outgrowths of the lacrimal bone) whose function is unknown. It has been suggested that these horns, which were located above and in front of each eye, were for display purposes, and possibly anchored a decorative crest.

It is obvious that *Allosaurus* was a large, fierce predator, but it is unknown exactly what this dinosaur ate. Many scientists suggest that *Allosaurus* preyed on medium-sized sauropods, or larger individuals that were sick or injured. This notion is supported by a February, 2001 analysis by Emily Rayfield, which suggested that *Allosaurus* was well-adapted to taking a quick downward bite into its victim, a hunting method that is well-suited for bringing down defenseless or juvenile animals. Some sauropod bones from Utah reportedly show *Allosaurus* bite marks. Based on fossil evidence,

some paleontologists have suggested that *Allosaurus* may have hunted in groups, which would have made it easier to ambush larger prey. It has even been suggested that *Allosaurus* was a scavenger, but this hypothesis is lacking in evidence.

However, despite this confusion over the hunting habits of *Allosaurus*, the discovery of this ancient carnivore is well-documented. The first dinosaur to be discovered in Utah was found during an 1859 U.S. Army Corps of Topographical Engineers expedition led by Captain John H. Macomb. A number of bones from the front leg of a giant animal were collected by the expedition's scientist, Dr. John S. Newburry, and were later named *Dystrophaeus viaemalae* by E. D. Cope. Today this dinosaur is known as a sauropod, but the find is important to the discovery of *Allosaurus*. This sauropod intrigued Cope so much that he hired fossil collectors to comb the Utah-Colorado area. In 1877, in a short paper, Cope mentioned and named *Allosaurus* based on a specimen uncovered in 1869. It was not until 1883 that one of his collectors, M. P. Felch, discovered the first virtually-complete *Allosaurus* skeleton, in Colorado. This skeleton was only the beginning, as countless more would eventually turn up.

While Felch's Colorado specimen was noteworthy, it paled in comparison with a 1927 discovery in Utah. During the summer of that year, a local sheepherder stumbled across a large black bone lying on the ground near Cottonwood Wash in Emery County. Knowing that the bone did not belong to a sheep or any other modern animal, the herder told his friends. It didn't take long until word reached the academic community. Only two years later, in 1929, the first official dig was conducted at the Emery County site. This excavation, conducted by the University of Utah, uncovered nearly 500 dinosaur bones during the first season of work—and better things were yet to come.

In 1939, a Princeton student named William Lee Stokes became intrigued by the site, and mounted a small expedition. Bringing only a few shovels and a poorly-paid assistant, Stokes began work. Soon, his expedition received financial support from Malcomb Lloyd, and began to turn up more specimens. Because of its close proximity to Cleveland, Utah and the donations of Lloyd, the site soon became known as the Cleveland-Lloyd Dinosaur Quarry, a name familiar to dinosaur lovers of all ages. Years after the Stokes excavation, in 1960, the University of Utah embarked upon a five-year project. The university hired Stokes to lead the project, with assistance from a young researcher named James H. Madsen. Two metal buildings were erected over the quarry in 1976, and now protect the exposed bonebed from weather and vandals. Today, the site can be visited by both researchers and tourists, and continues to turn out good specimens.

Since the initial excavation, more than 15,000 fossil bones have been removed from the site, representing at least 70 different animals and 14 Jurassic dinosaur species. Among these specimens, it is estimated that at least 44 are *Allosaurus* individuals. Why the glut of carnivore bones? It has been suggested that the site was once a groundwater-fed wetland, and over time became a trap where animals would get stuck in the mud. Predators, such as *Allosaurus*, would then attempt to eat the entrapped animals, only to become mired themselves. Although this sounds tragic, this ancient drama is a boon for paleontologists. Along with *Allosaurus*, good specimens of *Camptosaurus*, *Stegosaurus*, and *Ceratosaurus* have also been found at the site. It is even believed that an egg found at Cleveland-Lloyd may be that of an *Allosaurus*.

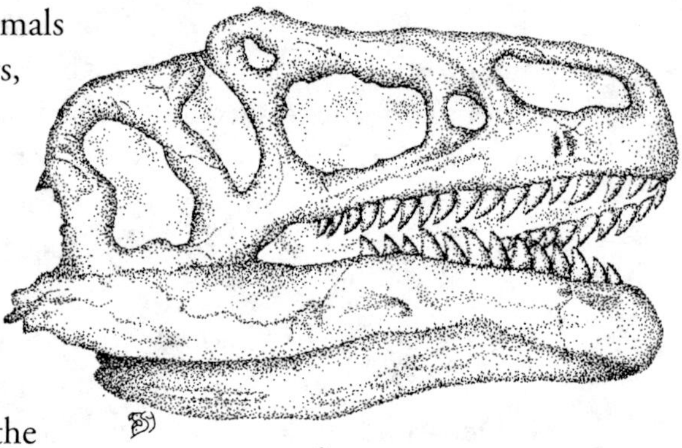

Allosaurus Fragilis ©Shannon S. Yeager, 2000!

Despite these amazing discoveries, it wasn't until 1988 that *Allosaurus* became the official Utah state fossil. Credit for this designation goes to one of the Cleveland-Lloyd paleontologists, James Madsen, who in 1979 recommended that *Allosaurus* be given the title. The Division of State History took up the cause, suggested other candidates (including a trilobite and a short-faced bear), and then invited elementary and junior high students to cast ballots for their favorites at the Utah State Fair. *Allosaurus* won. Later, in 1988, the State Legislature unanimously passed a measure officially making *Allosaurus* the state fossil. Luckily for Utah residents and prehistory fans, good *Allosaurus* fossils can be seen at several state museums, including the Utah Museum of Natural History, Earth Science Museum at Brigham Young University in Provo, the

College of Eastern Utah Prehistoric Museum in Price, the Museum of Natural Science at Weber State University in Ogden, the Museum of the San Rafael in Castle Dale, Dinosaur National Monument in Jensen, the Dinosaur Museum in Blanding, and, of course, the Cleveland-Lloyd quarry visitors' center.

While the whole of the American west is a haven for fossils hunters, it is Utah that possesses the most diverse Mesozoic fossil record of any American state. Of the plethora of dinosaurs known from Utah, *Allosaurus* is the most common, well-known, and popular. This Jurassic theropod is a perfect state fossil.

FOR FURTHER INFORMATION:

Averett, W.R., 1987. Paleontology and Geology of the Dinosaur Triangle. Museum of Western Colorado, Grand Junction.

—, 1991. Guidebook for Dinosaur Quarries and Tracksite Tour, Western Colorado and Eastern Utah. Museum of Western Colorado, Grand Junction.

Boren, R., 1998. *Allosaurus* Adopted Utah as Stomping, Chomping Ground. Deseret News, April 30, 1998, p. C02.

Brusatte, S, 2000. Kids' Corner-*Allosaurus*. Fossil News: Journal of Avocational Paleontology, vol. 6, no. 5, p. 11.

Chure, D.J., and Madsen, J.H., (in press). *Allosaurus*: The Making of Utah's State Fossil. in Gillette, D.D., (ed.), Vertebrate Fossils of Utah, Utah Geological Survey Publication.

Hayden, M., Dinosaurs of Utah, Utah History Encyclopedia, http://www.media.utah.edu/UHE/d/DINOSAURSOFUT.html

Madsen, J.H., 1976. *Allosaurus fragilis*: A Revised Osteology. Utah Geological Mining Survey Bulletin, 109: 1-163.

Stokes, W.L., 1986. Geology of Utah. Utah Museum of Natural History and Utah Geological Survey, Salt Lake City.

Turner, C.E., and Peterson, F., Jurassic Morrison Formation Ecosystem, National Park Service, http://www.aqd.nps.gov/grd/hot_topic/ht_morrison.htm

Cleveland Lloyd Dinosaur Quarry, Bureau of Land Management, http://www.blm.gov/utah/price/quarry.htm

Dinosaur National Monument, National Park Service, http://www.nps.gov/dino/

VERMONT

DELPHINAPTERUS

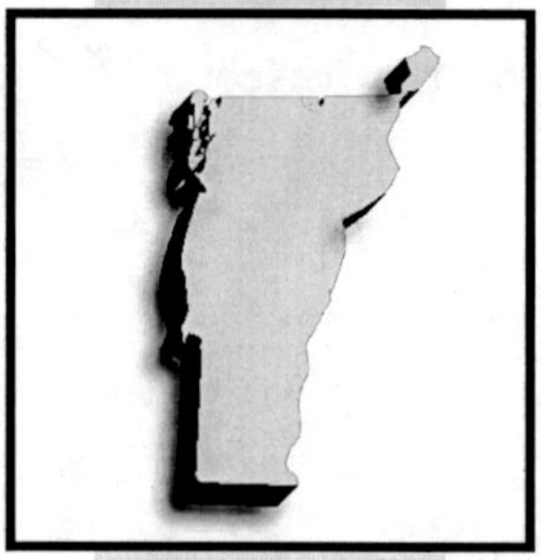

Vermont is a small state, both geographically and population-wise, and is composed largely of granite, which spells doom for fossil collecting. However, this quaint New England paradise boasts what may be the most peculiar of state fossils, *Delphinapterus leucas*. *Delphinapterus* is not strange in the sense that it possessed bizarre features or is found in only a small region, nor is it ancient or pivotal in the study of animal evolution. No, *Delphinapterus* is different from most state fossils because this particular genus and species is still living!

The odd story of the only extant state fossil begins about 12,500 years ago, during the waning days of the last great glaciation. During this time period, when prehistoric Cro-Magnon man was creating extravagant cave paintings in France, modern-day New England was covered by glaciers. In fact, the mass of ice was such that parts of Vermont sank under the weight of the glaciers, causing the land to be pushed lower than sea level. When the last great ice sheet melted away about 12,500 years ago, the sea poured into this depressed land. For about 2,500 years, until roughly 10,000 years ago, this portion of modern-day Vermont and New York existed as an arm of the Atlantic Ocean known as the Champlain Sea, which made up the southern boundary of what were the arctic regions of that time. This sea provided a habitat for molluscs, sea urchins, squid, fish, seals, and beluga whales, known to science as *Delphinapterus*.

Because *Delphinapterus* is an extant taxon, scientists know much about its physical characteristics, its social behaviors, and its evolutionary relationships. Therefore, *Delphinapterus* is probably the best-understood of the state fossils. While paleontologists know that this whale is recognized by a prominent forehead knob, known as a melon, and the absence of a dorsal fin, modern biologists are able to deduce much more about *Delphinapterus* by simply observing the animal in its environment. For instance, it is known that this whale is grey-white in color, is social, and uses echolocation to hunt. These characteristics are impossible to deduce from fossils alone.

Delphinapterus is a beluga, or white whale. Belugas, as mentioned above, are grey-white in color, with the grey gradually disappearing as the animal matures. By the time the whale reaches six years old, its skin is entirely white. Belugas are one of three whales that spend all their lives in arctic waters, and are special among the cetacean lineage as the only whales to be able to turn their heads, a functionality that possibly aids in communication. Beluga whales are also among the most sociable of aquatic mammals, and communicate using a wide variety of sounds. A group of belugas can be so loud that they are often nicknamed "sea canaries." In addition to communication, belugas also emit sound to hunt, a practice called echolocation. However, this hunting method, used to catch fish, pales in comparison with another beluga trait: assisted live birth. It has been reported by native people and some scientists that beluga whales actually help each other give birth! Belugas, unlike some other whales, have good vision, but lack a dorsal fin. This lack of a fin has earned belugas the generic name *Delphinapterus*, or "dolphin without a wing."

Although it is still around, *Delphinapterus* is a relatively rare animal today. Much of this is due to two factors: the gradual warming of the earth and modern-day pollution. Beluga populations declined as a result of glacial melting. As worldwide warming continued, the glacier covering the north shore of the Champlain Sea receded northwards. Free of the heavy glacier, the depressed lands of Vermont and New York rose. The Atlantic Ocean retreated eastward, causing the Champlain Sea to dry up and its outlines to shrink. The habitat of cold-water animals, including the beluga whale, withdrew northwards. Today, the remaining population permanently inhabits a marine arctic microenvironment, which is a short segment of the St. Lawrence Estuary centered at the

mouth of the Saguenay River. Because of the loss of habitat, coupled with the effects of pollution, the modern-day population of belugas comprises some 650 individuals, a number that garnered a status of an endangered population from the Canadian government in 1980.

The most important and intriguing *Delphinapterus* fossil ever discovered was nothing more than an accident. In August of 1849, while constructing the first railroad between Rutland and Burlington, workers unearthed the bones of a mysterious animal near the town of Charlotte. Found about 10 feet below the surface in a thick layer of blue clay, the workmen only paused momentarily to examine the bones before discarding them as the remains of a horse. Several days later, while walking near the construction site, Charlotte resident John G. Thorp observed several bone fragments in the dirt. These fragments intrigued him, and Thorp proceeded to convince the overseer to move the work to another segment of the project to allow for collection and study of these bones.

The man called in to study these bones was noted naturalist Zadock Thompson of the University of Vermont. Thompson returned to the site, collected the bones, and took them back to his lab. He determined that the greater part of the head, all of the teeth, and several vertebrae, ribs, and limb bones were undamaged by the workmen, and, therefore, able to be studied.

However, Thompson had little idea of what the bones represented.

Finding himself stuck, he began reading Georges Cuvier's classic 1825 comparative study on fossil bones, and only after reading this volume did he determine that the bones bore a strong resemblance to *Delphinus** *leucas,* the extant white whale. Thompson later wrote of his discovery, "My next objective was to ascertain, if possible, whether it was a living or extinct species of (the whale) family. Being without specimens for comparison, my only reliance for aid was Cuvier's great work on fossil bones. By a comparison of the fossil whale with the descriptions and figures in that work, it was found to resemble the living rather than the extinct types, and that the (structure) of the head was very (much) like that of the beluga *leucas*, or small northern White Whale."

However, despite the resemblances, Thompson proposed a provisional name: *Delphinus vermontanus*, paying homage to the state of Vermont. Thompson's actions were later reviewed in a 1908 paper written by Vermont State Geologist George H. Perkins, who wrote, "the generic status of the specimen is correctly determined…but whether it is identical with the living species *leucas* or should be placed as a new species, as Thompson thought, is not so readily decided." Only years after both scientists studied this specimen was it determined that the name *Delphinapterus leucas* should also describe the Charlotte specimen. Finally, nearly 125 years after the original discovery of the Charlotte whale, the Vermont state legislature made *Delphinapterus leucas* the state fossil with the 1993 passage of Act No. 66.

Despite the controversy over its scientific name, scientists do know that the Charlotte whale, nicknamed "the Vermont whale," died approximately 11,000 years ago of unknown causes. Its body eventually came to rest at the bottom of a quiet bay or estuary in the Champlain Sea, undisturbed by predators or scavengers. It was buried along with an assemblage of clams, mussels, and plants that inhabited the same waters. The specimen is approximately 4 meters in length, and some scientists believe that it is a female. This deduction is based on the presence of fossil mollusc shells that inhabited only shallow water. Modern female belugas and their young generally enter shallow waters, while males do not. It is known that the Vermont whale is not a juvenile, but it

* The beluga was first classified as a dolphin, *Delphinus leucas,* in the late 18th century by German naturalist Peter Pallas—hence the generic name *Delphinus.* It was later given its own genus, *Delphinapterus.*

cannot be proven that it was indeed a female, as there is no way to know if it died in the shallow waters or was just buried there.

Delphinapterus is unique among state fossils in that its remains are not confined to mere museums. Although they are rare, modern, living beluga whales can be seen in some parts of Canada. However, for the conformist paleontology buff, the best places to see the fossil bones of *Delphinapterus* are at the University of Vermont and the Redpath Museum in Montreal. Of course, the cold waters of Canada are always beckoning biologists to observe the only extant state fossil.

FOR FURTHER INFORMATION:

Brusatte, Steve, 2001. A Journey Through Deep Time: the Paleontology of New England and Quebec. Fossil News: Journal of Avocational Paleontology, v. 7, no. 4, p. 4-9.

—, 2002. Kids' Corner: *Delphinapterus*. Fossil News: Journal of Avocational Paleontology, v. 8, no. 6, p. 14.

Harington, C.R. Marine Mammals of the Champlain Sea, and the problem of whales in Michigan. in: Gadd, N.R. (ed.), The Late Quaternary Development of the Champlain Sea Basin. Geological Association of Canada, Special Paper 35, p.225-240.

MacRae, A. Champlain Sea Fossils.
http://www.geo.ucalgary.ca/~macrae/t_origins/champlain/champlain.html

Martineau, D., Diseases and Causes of Death of Beluga from the Saint-Lawrence Estuary, Quebec, Canada. http://www.medvet.umontreal.ca/services/beluga/beluga_homepage.html

Perkins, G. H., 1908. Fossil Cetacea of the Pleistocene of the United States and Canada (with special reference to *Delphinapterus vermontanus*, Thompson). Report of the State Geologist of Vermont, 1907-1908, pages 76-112.

Thompson, Z., 1850. An account of some fossil bones found in Vermont, in making excavations for the Rutland and Burlington Railroad. American Journal of Science, volume 9, no. 26, March 1850, pages 256-63.

—, 1853. (Reference to both the Charlotte Whale and the Mt. Holly mammoth). Thompson's Vermont, Appendix, pages 15-20.

Vermont Secretary of State, Vermont Legislative Directory and State Manual, Biennial Session, 1993-1994, p. 24.

VIRGINIA

CHESAPECTEN JEFFERSONIUS

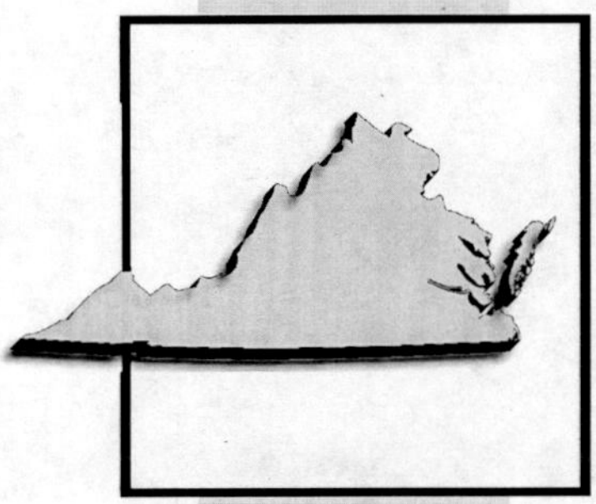

Most historians would agree that Thomas Jefferson is one of the most revered men in American history. A President, inventor, naturalist, architect, writer, and thinker, Jefferson embodied the multi-talented early American. However, few realize that Jefferson was a fossil collector, amateur paleontologist, and even named extinct animals based on fossil evidence. It is no surprise, then, that the official state fossil of Jefferson's home state, Virginia, bears his name.

Although Jefferson documented many of the first fossils found in Virginia in a series of scientific papers and books, he was not the first to discover, illustrate, or describe *Chesapecten jeffersonius*, the state fossil named after him. Instead, this honor goes to the first people to establish a permanent English colony in America—the settlers of Jamestown. Although reports vary, it has been written that during the first weeks after Jamestown was settled, the colonists noted Native Americans using bowls and scraping tools made out of what appeared to be shells. A band of colonists then reportedly explored the local rock units, and discovered abundant fossil shells. Many of these specimens were sent back to London. While there is no proof of this story, *Chesapecten* fossils are commonly found in the Jamestown area.

Exactly what was this fossil that so greatly intrigued America's first permanent English settlers? *Chesapecten*

is a scallop, and therefore a mollusc and member of the Family Pectinidae. Like all scallops, *Chesapecten* was a bivalve. However, unlike other molluscs such as clams, the shells of scallops, including *Chesapecten*, sport heavy ridges, or ribs. These ribs make the shells stronger, helping to protect the scallop from predators such as fish. Like oysters and clams, scallops typically lie on the seafloor with one shell up and one down. When the scallop relaxes on the seafloor, the two shells also open slightly, exposing a fold of mantle tissue called the velar fold. This fold protects numerous tentacles, used for feeding, that are sensitive to stimuli.

In addition to the ridges, scallops also differ from other bivalves because of their eyes and the ability to swim. Along both edges of the mantle are numerous iridescent ocelli, or primitive eyes. These eyes are sensitive to decreases in light, such as those produced by passing shadows. This adaptation helps the scallop avoid predators, as does the ability to swim. When the scallop claps its valves, water jets out through gaps in the shells, propelling the animal forward. Although this swimming method is rudimentary, it is a major defensive weapon.

Modern scallops are normally found in the shallow waters of protected bays with sandy or muddy bottoms. Because of this, it is expected that *Chesapecten jeffersonius* lived in much the same type of habitat. In fact, most paleontologists believe that *Chesapecten* lived along the continental shelf at shallow to moderate depths.

Chesapecten fossils are found from New Jersey to Florida, but are especially common in the Chesapeake Bay area of Virginia and Maryland. Most specimens, which average 6-9 inches in width, are found in Late Miocene and Pliocene sediments, and are dated to between 4 and 5 million years ago. While some *Chesapecten* species survived to more modern times, *C. jeffersonius* likely became extinct about four million years ago. Exactly why did it become extinct? During the end of the Pliocene the climate became cooler. *Chesapecten*, and other molluscs, may have become extinct because they could not adapt to this temperature change. Or, possibly, a scallop-eating predator, such as a new fish, may have appeared. The answer is a mystery.

Regardless of how *Chesapecten jeffersonius* became extinct, it is obvious that this bivalve has captivated the minds of both early and modern American scientists. In fact, *C. jeffersonius* has the distinction of being the first fossil from North America to be illustrated in a scientific publication, appearing in Martin Lister's 1687 book *Historiae Conchyliorum* (History of Conchs). Seven years later Lister described *Chesapecten jeffersonius* in a different publication, but it was not until nearly a century and a half later that it was officially named, in 1824.

Since its original discovery and description, *Chesapecten* has helped scientists reconstruct the Pliocene environment of Virginia, and learn more about the evolution and extinction of prehistoric bivalves. Since the time of its discovery, literally thousands of *Chesapecten* specimens have been collected. However, this nearly 400-year yield of fossils was surpassed by the actions of one monster storm: Hurricane Floyd.

Floyd, which struck the Virginia coast during the waning months of 1999, was one of the strongest and most dangerous hurricanes to hit the American south in recent history. Its strong winds and rains caused millions of dollars' worth of damage, but exposed just as many millions of fossils. Many of these exposed fossils were specimens of *Chesapecten jeffersonius*. Most of these *Chesapecten* fossils were discovered embedded in the walls of a 30-foot-wide ditch cut into a Williamsburg golf course. This ditch, which cut through Late Miocene sediments, was created in a period of three days, when water rushing from nearby Lake Matoaka blasted away a large area of the grounds. Although unfortunate for golfers, the ditch was a boon to geologists.

Among those studying the golf course was Gerald Johnson, a College of William and Mary professor and expert on Miocene and Pliocene invertebrates. In the days after Hurricane Floyd subsided, Johnson received several calls alerting him to a number of newly-discovered fossil outcrops. The golf course ditch was only one of these. However, Johnson said the ditch was one of the most spectacular, and the best *Chesapecten*-bearing outcrop he had seen in his 35-year career. As Johnson whimsically told a reporter after the golf course discovery, "a little hurricane does a lot for geology."

However, this golf course is not the only place to find good *Chesapecten* fossils. In fact, there are several sites in Virginia where one can legally collect these Tertiary bivalves. All of these sites are located on the Virginia Coastal Plain, which extends from the state's mountainous Fall Zone eastward to the Atlantic Ocean. The Coastal Plain is a terraced landscape that literally stair-steps down to the coast. These terraces are remnants of former shorelines, including the Miocene-Pliocene sea in which *Chesapecten* lived. As would be expected, the Coastal Plain is underlain by a thick layer of sediments, including both Miocene and Pliocene rocks. These rock layers are a bonanza for fossil hunters.

One of the best collecting localities on the Coastal Plain is in York County, located at the mouth of Indian Field Creek and along the south bank of the York River. Late Miocene sediments are exposed at this site, and fossils of several fish, bird, reptile, and mammal species coexist with amazing bivalve specimens. *Chesapecten* is one of the most common of these fossilized bivalves.

The York County site is a microcosm of the entire state of Virginia: a diverse fossil record highlighted by wonderful bivalve molluscs. One of the most common and unique of these bivalves is *Chesapecten*. Because of its ubiquity and uniqueness, the Virginia General Assembly designated *Chesapecten jeffersonius* the official state fossil in 1993. This designation respectfully honors a great fossil and a great genius and naturalist, Thomas Jefferson.

FOR FURTHER INFORMATION:

Anonymous. Fossils, thanks to Floyd: Hurricane helps uncover outcrops. Associated Press, September 27, 1999.

Brusatte, S., 2002. Kids' Corner-*Chesapecten*. Fossil News: Journal of Avocational Paleontology, v. 8, no. 3, p.15.

Campbell, L.D., 1993. Pliocene Molluscs from the Yorktown and Chowan River Formations in Virginia. Virginia Division of Mineral Resources Publication 127, 259 p., 43 plates.

Ray, C.E. (ed), 1987. Geology and Paleontology of the Lee Creek Mine, North Carolina, II. Smithsonian Contributions to Paleobiology No. 61, 283 p., 85 plates.

Ward, L.W. and Blackwelder, B.W., 1975. *Chesapecten*, a New Genus of Pectinidae (Mollusca: Bivalvia) from the Miocene and Pliocene of Eastern North America. US Geological Survey Professional Paper 861, 39 p., 7 plates.

College of William and Mary Department of Geology, Coastal Plain Province: The Geology of Virginia, http://www.wm.edu/CAS/GEOLOGY/virginia/coastal_plain.html

WASHINGTON

MAMMUTHUS COLUMBI

While *Tyrannosaurus rex* and other dinosaurs may be well-known to school children and trilobites and brachiopods are commonly found and studied by fossil collectors, it is the mammoth which may just be the most popular state fossil. Despite the fact that this large elephant has been extinct for only some 10,000 years, three different states have designated the mammoth as their official state fossil: Alaska, Nebraska, and Washington. However, the fossil mammoths designated by each state, although very similar, are also different.

Nebraska became one of the first states to designate a state fossil by recognizing the generic "mammoth" in 1967. Despite the ambiguity in the designation, the state fossil of Nebraska really refers to *Mammuthus imperator,* an animal well-known from a famous 1920s discovery. Alaska, on the other hand, recognized *Mammuthus primigenius*, or the woolly mammoth, one of the best-known prehistoric creatures. Washington, the most recent state to name a mammoth as its state fossil, chose *Mammuthus columbi*, or the Columbian mammoth, another well-known prehistoric elephant.

Like other mammoths, the Columbian can trace its ancestry to *Mammuthus meridionalis*, an early species that crossed the Bering Land Bridge and diversified into the wide variety of mammoths now known from North America. And, like other mammoths, the Columbian mammoth lived during the Pleistocene, some 1.6 million-10,000 years ago. However, despite their common ancestry, the woolly mammoth, imperial mammoth, and Columbian mammoth are all quite different. While the woolly mammoth was restricted to colder, more northern habitats, the Columbian mammoth was more cosmopolitan, and ranged from Alaska, across the United States, and down into Mexico and Central America. Based on fossil remains, it is likely that this species favored grasslands. Unfortunately for the grasslands, the Columbian mammoth was a food-processing machine.

A Columbian mammoth individual was huge, to put it bluntly, standing nearly 13 feet tall at the shoulder and weighing 8-10 tons. It is likely that the Columbian mammoth could consume about 700 pounds of vegetation a day. Multiply this by its 60-80 year lifespan, and that is a lot of vegetation! Obviously, the mammoth had to evolve certain features and structures to help it consume this much grass. Two of these structures, teeth and tusks, are among the most successful grazing adaptations ever seen in fossil animals.

Mammoth molar teeth, common fossils in Washington, are fascinating. Along with simply aiding the mammoth in feeding, these teeth can assist scientists by telling them exactly how old the mammoth was when it died and which species it belonged to. Like their modern elephant cousins, mammoths possessed six sets of teeth throughout their lifespan, with their last and largest set consisting of four shoebox-sized tooth plates. These large molar teeth were covered in ridges, which helped the mammoth break apart tough vegetation. By the age of six, the mammoth had already acquired its first three sets of teeth. By age 13, the fourth set had arrived, and by age 27 the fifth set had pushed its way through the mammoth's jaw. The final set would

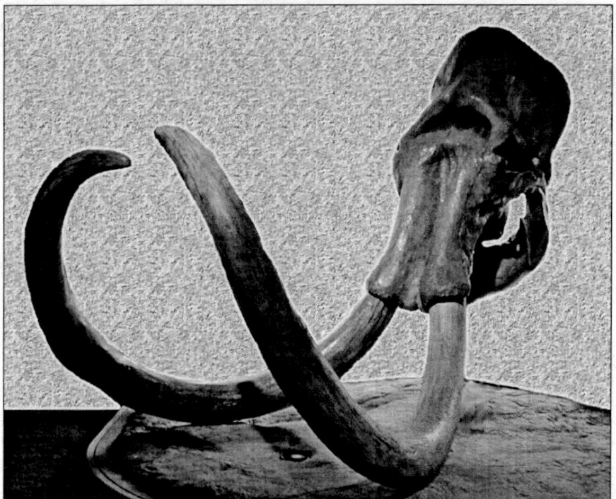

arrive at approximately age 43, and lasted for the remainder of the mammoth's lifetime. When this last set wore away, the elderly mammoth would die because of its inability to feed.

Researchers can use these teeth to determine the approximate age of the mammoth by measuring their length and width, and comparing them to measurements from modern-day elephants. However, the second major feeding adaptation, the tusks, are much better for determining age. Unbeknownst to many, the tusks of mammoths and other elephants are actually teeth themselves—enlarged and elongated incisors. These teeth begin growing at birth and continue throughout the life of the mammoth. A thin layer of enamel is present initially at the tip of the tusk, but quickly wears away. The majority of these tusks are composed of dentine, which is deposited in thin circumferential layers. By counting these layers, which are deposited annually, scientists can determine the age of the mammoth when it died—sort of like tree-ring dating.

The tusks were likely primarily used for feeding, but probably also had a variety of functions. Most scientists believe that they had at least some role in digging up vegetation, a hypothesis supported by the behavior of modern elephants. A study conducted by scientists from the Mammoth Site in South Dakota seems to show that some mammoths were "right-tusked" or "left-tusked," much like humans are right- or left-handed. Perhaps, scientists have speculated, the mammoth may have favored one tusk when digging up vegetation. However, many scientists also believe the tusks of mammoths—and modern elephants—have several more functions, including intraspecies fighting, snow "plowing," deterring predators, and sexual display. Regardless, these tusks, which often grew to lengths of over 10 feet, were definitely used for something!

While mammoths are now extinct, fossils of the Columbian mammoth do frequently turn up in Washington and several other states. In fact, there are about 40 documented mammoth discoveries in Washington alone. Some of the most famous Columbian mammoth fossils found in Washington were discovered in 1996, when a bonanza of bones was uncovered on the Olympic Peninsula near Sequim. In recent years tusks have been discovered near Lakeview Elementary School in Kirkland, and on May 10, 2001, a strange tusk was discovered in the small town of Moxee, near Yakima. This 4.5 foot tusk was discovered when Steve Herke was grading a new parking lot. The Yakima Valley Museum will excavate the tusk, prepare it, and make it part of an exhibition on the Yakima region's prehistoric past.

However, long before the recent discoveries, a Washington high school class became interested in sponsoring a state fossil bill. After consulting with a paleontologist, they picked five candidates: two fossil crabs, a fossil palmetto, a red cedar, and the giant ground sloth. They finally settled on the sloth, but the State Legislature sunk the bill. In 1994, however, a different group of students, this time from Windsor Elementary School near Cheney, began a four-year effort to have the Columbian mammoth named the state fossil. Their efforts culminated in the 1998 designation of this well-known mammoth.

While the Columbian mammoth may not be as well-known as the woolly mammoth, this prehistoric elephant was an important part of the Ice Age ecosystem of Washington. Because of its importance and several recent finds in Washington, this pachyderm was designated the state fossil, a worthwhile honor for an animal only recently extinct.

FOR FURTHER INFORMATION:

Anonymous. Grading Unearths Mammoth Tusk, Everett Daily Herald, May 25, 2001.

Brusatte, S., 2002. Kids' Corner-mammoth. Fossil News: Journal of Avocational Paleontology, v. 8, no. 7, p.16-17.

Nash, J.M., 2001. Who Killed Woolly. Time Magazine, vol. 157, no. 24, p. 64-65.

Ward, P.D., 1997. The Call of Distant Mammoths. Copernicus Press, New York, 241 p.

Burke Museum of Natural History, Fossil Vertebrates, http://depts.washington.edu/vertp/

Life and Times of Washington State, http://www.washington.edu/burkemuseum/ltws.html

Mammoth Site, Columbian and Woolly Mammoth Information, http://www.mammothsite.com/mammothinfo.html

North Bellingham School Facts Page, State Fossil, http://www.ferndale.wednet.edu/nb/facts/state/fossil.html

WISCONSIN

CALYMENE CELEBRA

Over the course of geologic time, Wisconsin has been home to billions of intriguing life forms, from early arthropods to dinosaurs to the recent mammoth and musk ox, which only yesterday disappeared from the state's rolling hills and fertile plains. Unfortunately, the most recent Ice Age, including the major glaciation which bears Wisconsin's name, wreaked havoc on the Dairy State's fossil record. Remaining surface exposures in the central and northern part of the state are nearly barren of fossils, instead being composed of thick layers of glacial till. But, the area of Wisconsin extending from Prairie du Chien in the west to the urban center of Milwaukee on the shores of Lake Michigan is chock full of fossils, including a magnificent trilobite now designated the state fossil.

The trilobite in question, *Calymene celebra*, enjoyed unparalleled success when shallow, salty seas covered much of Wisconsin during the Late Ordovician and Silurian. During this time Wisconsin was about 30° south of the equator, a position which encouraged the growth of coral reefs. These reefs, especially prevalent in Silurian deposits, provided a haven for brachiopods, corals, crinoids, and, of course, several species of trilobites.

Calymene celebra was one of the most numerous of these species. Like all trilobites, *Calymene* was an arthropod, and had jointed legs, an exoskeleton, and a segmented body. Its exoskeleton was divided into three basic transverse sections—the cephalon, which bore eyes and antennae; the thorax; and the pygidum, or tail. However, *Calymene* can be distinguished from other trilobites by its near-triangular cephalon, which contains a large glabella, or circular node. Furthermore, this glabella was surrounded by a deep furrow and four rounded extensions. The posterior margin of the cephalon forms the widest part of the exoskeleton, and is easily seen in fossil specimens. The thorax of *Calymene* contained 12-13 segments, while the pygidum bore six. In addition, the pygidum of *Calymene* tapers rapidly, but has a morphology that is otherwise similar to that of the thorax.

Calymene celebra grew to a length of about ten centimeters (4 inches), but most specimens are much smaller. Many *Calymene* fossils are discovered in rocks that indicate shallow-water, continental-shelf deposition, especially limestones and dolomites, while several species have been discovered in reef environments, including *Calymene celebra*. The presence of reefs may indicate that some species of *Calymene* preferred warm, shallow environments, although many species have also been found in deep water deposits. Fossil evidence shows that *Calymene* likely lived on the seafloor, where it crept or crawled in search of food.

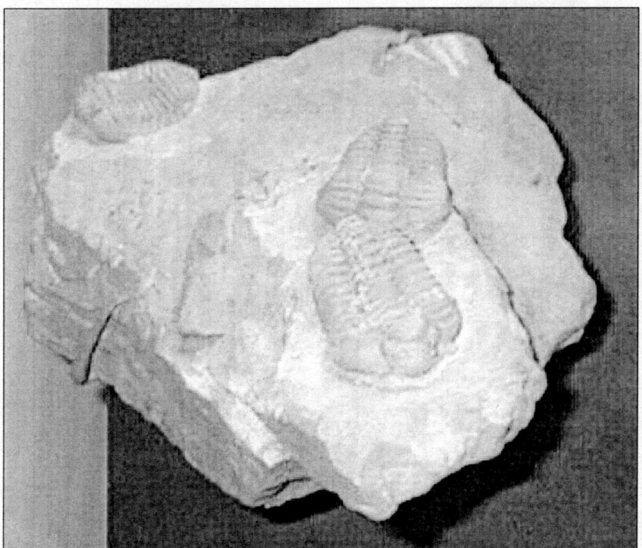

Calymene fossils have offered paleontologists a glimpse into predator-prey relationships of the Silurian. The final 18-19 appendages of *Calymene celebra* and other members of the genus bore sensory bristles, many of which have been preserved in Wisconsin specimens. These bristles enabled *Calymene* to roll itself into

a ball, much like a modern pill bug, which allowed the trilobite to protect its soft underbelly from predators. Many Wisconsin specimens have been discovered rolled up into a ball.

As a genus, *Calymene* first evolved in the Late Ordovician or Early Silurian. In Wisconsin, it attained maximum diversity during the Silurian, when it was a major part of the reef fauna. However, as inland seas began to dry up during the Devonian, *Calymene* became extinct both in Wisconsin and throughout the world. While the exact reason for this extinction is unknown, either the evolution of a new predator or a period of environmental change likely led to the downfall of this trilobite.

Calymene remains one of the most important fossils found in Wisconsin. The first *Calymene* fossils were collected by naturalist Increase A. Lapham, who frequently hunted Wisconsin trilobites in the 1830s and 40s. Since that time, *Calymene* has become valuable as an index fossil, and is often used to correlate Silurian rock units in Wisconsin and other parts of the Midwest.

Because of its importance, many rockhounds and scientists lobbied to make *Calymene celebra* the official state fossil of Wisconsin. In 1982, University of Wisconsin-Milwaukee geology buff Mark Shurilla waged a campaign to adopt the "trilobite" as the official state fossil. However, University of Wisconsin-Madison geology professor Klaus Westphal was against the idea of simply designating the "trilobite." Instead, he suggested *Calymene celebra.*

Later, a *Calymene* bill was introduced in the legislature, but State Representative Joanne Duren suggested that the mastodon receive the official title. After some pressure from *Calymene* supporters, Representative Duren suggested that the mastodon be designated the official "vertebrate fossil," and the trilobite the official "invertebrate fossil." This bill passed the House, but died in the Senate.

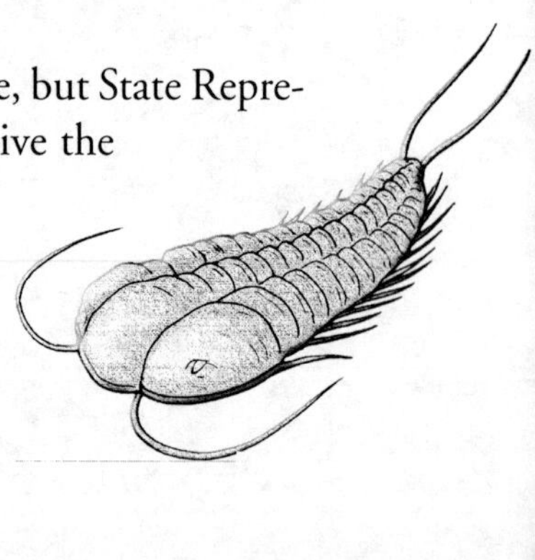

One year later, in 1983, the trilobite bill was reintroduced. This time, the legislation was backed by the Wisconsin Geological Society. A member of the society had a daughter, Jeannette Bell, who had just recently been elected to the State Assembly. Representative Bell sponsored the trilobite bill, which was then supported by Illinois State Geological Society paleontologist Donald Mikulic, a University of Wisconsin graduate. However, the trilobite bill died again, thanks in part to what Westphal referred to as "dirty political tricks."

The trilobite bill was introduced for a third time in 1985. This time, members of the Wisconsin Geological Society were more politically savvy, as President Margaret Pearson turned the adoption of the trilobite into a statewide issue, enlisting the support of several rockhound groups along with the University of Wisconsin and the Milwaukee Public Museum. Her work paid off when Governor Early signed Act 162, Amended Section 1.10, which officially designated the trilobite *Calymene celebra* as the Wisconsin state fossil on April 2, 1986.

Good specimens of *Calymene* can still be found in Wisconsin, along with various other species of trilobites. Most of the better-preserved *Calymene celebra* fossils are found in the state's vast Niagara Dolomite Formation. Several Silurian reefs are preserved in this formation, with one of the largest being the Wauwatosa Reef. This reef outcrops on the northeast corner of 68th and State Streets in Wauwatosa, and is some 19 meters thick and over 140 meters in breadth. Eighty-nine fossil species have been found at this reef, including *Calymene.*

In addition, collectors can find well-preserved trilobites in Ozaukee County, along the banks of the Milwaukee River between the cities of Cedarsburg and Grafton. Other specimens can be found in Milwaukee County, along the Menomonee River near the aforementioned Wauwatosa. Smaller collecting areas are also present in Racine and Waukesha Counties.

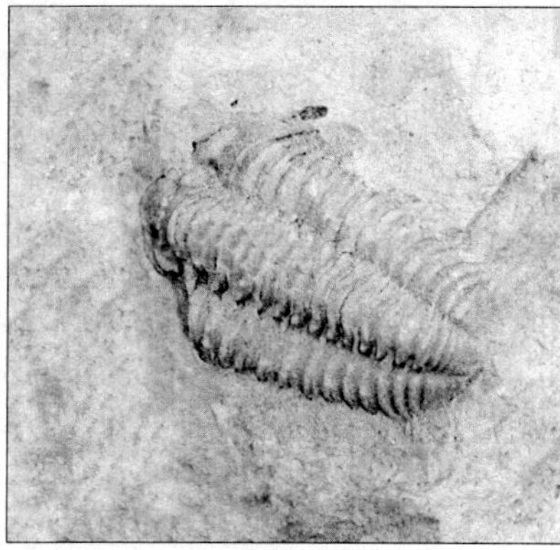

Luckily, *Calymene celebra is* a well-known fossil, and specimens are widespread in Wisconsin's Silurian rock units. Unfortunately, the same cannot be said of many other fossils, as the most recent Ice Age has done a number on the paleontological record of Wisconsin. Regardless, this small trilobite stands tall, as it has survived both the Ice Age and dirty political tricks.

FOR FURTHER INFORMATION:

Brusatte, S., 2000. Kids' Corner-*Calymene celebra.* Fossil News: Journal of Avocational Paleontology, v. 6, no. 10, p. 14.

Copper, P. and Brunton, F. 1991. A global review of Silurian reefs. Special Papers in Palaeontology, v. 44, p. 225-259.

Hall, J., 1867. Descriptions of some new fossils from the Niagara Group. New York State Cabinet of Natural History, 20th Annual Report, p. 308-401.

Lowenstam, H.A., 1957. Niagaran reefs of the Great Lakes area. Geological Society of America Memoir 67, p. 215-248.

Mikulic, D.G., Briggs, D.E.G. and Kluessendorf, J., 1985. A new exceptionally preserved biota from the Lower Silurian of Wisconsin, U.S.A. Philosophical Transactions of the Royal Society of London, v. B311, p. 75-85.

Ostrom, M.E., 1961. Fossil Collecting in Wisconsin. Wisconsin Geological Survey.

Watkins, R., 1993. The Silurian (Wenlockian) reef fauna of southeastern Wisconsin. Palaios, v. 8, p. 325-338.

Environmental Education for Kids, Wisconsin State Fossil, http://www.dnr.state.wi.us/org/caer/ce/eek/nature/state/fossil.htm

Milwaukee Public Museum, Virtual Silurian Reef, http://www.mpm.edu/reef/ww-body.html

WYOMING

KNIGHTIA

Millions, perhaps billions, of fossils are housed in museums and personal collections, coming from every corner of the earth. But only a handful of fossil localities can be classified as "lagerstatten" or "super sites," places where an abundance of well-preserved fossils from one geologic age are found—Solnhofen, Mazon Creek, and Ghost Ranch immediately come to mind. Another gold mine is the Green River Formation of Wyoming, Utah, and Colorado, a large unit of Eocene rock that preserves a glut of fossil fish. Of these fish, the herring *Knightia* is the most common, and is today recognized as Wyoming's official state fossil.

The Green River Formation can trace its origin back to the Cretaceous Period, when a prolonged period of tectonic movements ripped through the nascent Rocky Mountains, creating a series of large basins. Later, during the Eocene, numerous streams and rivers flowed into these basins and fed lakes, many of which existed for only a short time geologically, but were still home to countless fish species. This formation consists of three distinct units: the Uinta Basin in Utah, the Piceance Basin in Colorado, and the Green River Basin in Wyoming. It is the Green River Basin of Wyoming that is the most dear to fossil collectors.

This basin preserves the faunas of a large lake system, composed of three distinct lakes and several smaller ones. These three lakes, now dubbed Fossil Lake, Lake Gosiute, and Lake Uinta, varied in size and depth through time. As a whole, this system existed for about 20 million years, forming about 60 million years ago and disappearing towards the end of the Eocene. For the duration of these 20 million years, what is now Wyoming was a subtropical land, and was the home of an interesting and distinct, yet modern-looking, fish population.

This fish population is well-understood from each of the three lakes. Lake Uinta was the largest in surface area of the three, but was the shallowest. It was also the longest-lived of the three lakes, existing for some 17 million years. However, its thick sediments, which reach 7,000 feet in some places, do not produce most of the fossil fish known from Wyoming. This distinction goes to the two remaining lakes: Gosiute and Fossil. Of these lakes, Gosiute was the most extensive. Based on the shape of the sediments, it is known that the level of Lake Gosiute constantly fluctuated, which led to the frequent drying out of thousands of square kilometers of lake floor. This desiccation led to the preservation of a high-grade oil shale, a rock rich

in fossils. However, the best-known and most recognized fish fossils do not come from Lake Gosiute, but from Fossil Lake, which was the deepest of the three. It had a small surface area and was relatively short-lived, but produces the largest share of fish specimens of the three lakes. Much of what remains of Fossil Lake is preserved at Fossil Butte National Monument, near Kemmerer, a popular site with tourists and paleontologists.

Among the fish fossils commonly found in Fossil Lake is *Knightia*. According to the Denver Museum of Nature and Science, *Knightia* is the most commonly collected vertebrate fossil in the world. This notion is supported by the fact that in 1978 alone more than 20,000 *Knightia* specimens were unearthed. Many of these specimens came from the Kemmerer area, where thousands of *Knightia* fossils are found, excavated, and commercially sold. In fact, these fish are so common that they are frequently seen in souvenir stores across the state.

In life, *Knightia* grew to a length of 5-10 inches, and was covered by layers of scales, similar to those seen in modern-day fish.

It was a member of the herring family, closely resembling extant herrings. *Knightia* possessed a deeply-forked tail and single dorsal fin in front of the mid-body line. It had a relatively large head covered with protective plates, and its vertebral column was located dorsally. Like many modern freshwater fish, it had four fins: one on its back, stomach, pelvis, and anus. *Knightia*'s diet likely consisted of plankton and algae, fossils of which have also been found in the Green River Formation. Based on fossil evidence, some larger fish also ate *Knightia*.

Because so many specimens are found together, it has been hypothesized that *Knightia* was a schooling fish. What caused the mass die-offs that resulted in these rich fossil deposits? The answer may be found under a microscope. When the remains of *Knightia* are viewed under high magnification, it can be seen that the fibers of the muscular tissue are frequently preserved intact. They no longer consist of muscle proteins, however, but of pure calcite. Because of this, the burial and preservation of these fish must have occurred in the absence of atmospheric oxygen and bacteria. It is likely, then, that low oxygen levels caused the vast schools to die, and their remains were quickly buried on the lake floor. Over time they turned to stone, where today several hundred individuals are often seen densely packed in one square meter of Green River rock!

Although it is likely that Native Americans first stumbled upon Green River fossils, missionaries began reporting strange fish fossils from Wyoming as early as the 1840s. In 1848, Jack Robinson, a fur trapper, and James Hall, an explorer, first wrote about fish fossils they discovered in Wyoming. In 1856, Dr. John Evans collected the first fossil fish from this formation known to science, some eight years after the initial report. Evans sent one of his specimens to Dr. Joseph Leidy, who named it *Knightia eocaena*, in honor of Wilbur Clinton Knight, the first Wyoming state geologist and a paleontologist at the University of Wyoming. Soon after, geologists, railroad workers, and missionaries started reporting fossil sites in surrounding areas. In 1868, the Union Pacific Railroad built the first rails through the Green River Formation, exposing a plethora of outcrops. This led to a decade-long mapping project, conducted from 1868-1878, in which Wyoming was first scientifically surveyed. This survey led to the discovery of several additional fish localities.

Most of the strange fish discovered after the survey were sent to Edward Drinker Cope, who described 30 new species between 1877 and 1884. Cope even travelled to Wyoming to collect specimens—a rare deviation for a man who usually relied on hired collectors to find and excavate specimens—and used his discoveries to publish a major work on the fishes of North America.

While collecting in Wyoming, Cope noted that several of his best specimens came from "near the main line of the Wasatch Mountains." Because of this ambiguous description, it is unclear where Cope discovered many of his better specimens. Regardless, fine fish fossils continued to turn up along Twin Creek, especially in an area dubbed "The Fish Cliffs," rock outcrops exposed on the west side of Twin Creek Valley. While many of these early sites have long since been closed, good specimens are commonly found in Wyoming today, and can frequently be collected, seen at museums, and bought in souvenir stores.

Knightia became Wyoming's official state fossil on February 18, 1987. Much of the credit for this designation is due to a young student at Anderson Elementary School in Cheyenne, Fred Hurlburt. While only in grade school Fred spoke in front of the State House of Representatives and Senate and frequently appeared on radio programs to promote House Bill 361, which called for the designation of *Knightia* as the official fossil and the cutthroat trout as the state fish. After his lobbying, Governor Mike Sullivan signed the bill into law.

The Green River Formation, because of its abundance of fish fossils, is definitely one of the most famous rock units in the world. Of these fish, *Knightia* is the most common, and one of the most frequently-collected vertebrate fossil specimens in the world. For these reasons, and many more, *Knightia* is the perfect state fossil for Wyoming.

FOR FURTHER INFORMATION:

Brusatte, S., 2000. Kids' Corner-*Knightia*. Fossil News: Journal of Avocational Paleontology, vol. 6, no. 8, p. 11.

Caggiano, T., 2000. The Green River Formation. Fossil News: Journal of Avocational Paleontology, vol. 6, no. 12, p. 4-7.

Grande, L., 1984. Paleontology of the Green River Formation, with a Review of the Fish Fauna. Geological Survey of Wyoming, Bulletin 63.

Long, J.A., and Young, G.C., 1995. Fossil Fishes. Chapman and Hall, London.

Pellant, C., 1996. An Illustrated Guide to Fossils. the Nature Company, Surrey, Great Britain.

The Green River Oil Shale-A Fossilized Energy Trap, http://www.amen.org.uk/vorwelt/megasucc/megac9.htm

STATE DINOSAURS

MARYLAND

ASTRODON JOHNSTONI

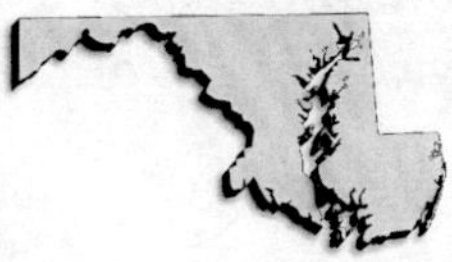

Maryland has designated both an official state fossil and an official state dinosaur. Maryland's state fossil, the snail *Ecphora gardnerae gardnerae*, is common in the state's Tertiary sediments, and is popular with collectors. On the other hand, the state's official dinosaur, the sauropod *Astrodon johnstoni*, is quite rare, poorly understood, and perhaps even without a proper scientific name.

Astrodon johnstoni lived 130-95 million years ago during the Early Cretaceous, a time during which Maryland sported a rich dinosaur fauna. The area in which *Astrodon* is found is widely considered to have been a swamp. However, it is unlikely that *Astrodon* actually lived in this swamp, but rather that it lived further inland, and its teeth and bones washed into the lowland via rivers after death.

Like most sauropods, *Astrodon* possessed a small head, long neck, and long tail. Judging from isolated fossil remains, this giant may have reached sizes of over 20 tons, and a length of 50-60 feet. *Astrodon* was an herbivore, and likely ate conifers, ferns, cycads, and other plants. A recent analysis shows that *Astrodon* could raise its head some 10 meters (33 feet) or more above the ground to feed on foliage that virtually none of its contemporaries could reach. While many sauropod teeth don't appear to be well-designed for chewing, the teeth of *Astrodon* often display wear at their tips. It is possible that this wear came from stripping off foliage from high branches and was unrelated to chewing, although this hypothesis is hard to test. Based on studies of similar sauropods, it has been suggested that *Astrodon* could run upwards of eight meters a second (18 miles an hour), although its normal gait was likely much slower.

The evolutionary relationships of this sauropod are reasonably well-known. *Astrodon* possesses at least four of the distinctive characteristics of the family Brachiosauridae, including long forelimbs and hands and pleurocoels in the vertebrae in front of the pelvis. Based on this information, *Astrodon* is a brachiosaurid, making it one of the few Early Cretaceous survivors of this otherwise Jurassic group.

The confusion over its proper scientific name has yet to be resolved. Throughout its history, *Astrodon* has been known as *Pleurocoelus*, and some scientists have suggested that it is synonymous with *Brachiosaurus*. It is currently not believed that *Astrodon* is a *Brachiosaurus* individual, but many do classify this Maryland sauropod as *Pleurocoelus*. However, this classification may be wrong. *Astrodon* was originally described based on an isolated tooth, a topic we shall visit in a moment. This tooth has some unique characteristics that are shared by only one other known sauropod: *Pleurocoelus nanus*, one of three named *Pleurocoelus* species and a dinosaur also known from Maryland. It is unlikely that two large, related sauropods of the same size lived in Maryland at the same time, so many suggest that *Pleurocoelus nanus* and *Astrodon johnstoni* are the same species. As the dinosaur named first, the name *Astrodon* receives preference over *Pleurocoelus*, making all *Pleurocoelus nanus* specimens actually those of *Astrodon johnstoni*. However, some have argued that the teeth on which the name *Astrodon* is based are not diagnostic, making *Astrodon* a nomen dubium.

The first *Astrodon* specimen, two isolated teeth, was discovered by Philip Tyson in November 1858. Tyson, then the State Agricultural Chemist for Maryland, discovered the specimens in an open-pit iron mine in the Arundel Clay on the property of John D. Latchford near Muirkirk, Prince George's County. He then showed the teeth to Dr. Christopher Johnston, a local physician and dentist, who sliced one very thinly for microscopic examination. Dr.

Johnston described the tooth in a paper published in the *American Journal of Dental Science* in 1859, and named the individual *Astrodon*, or "star tooth," a name referring to the appearance of the tooth's dentine in the cross section.

Despite Johnston's paper, *Astrodon* was still without a proper scientific description. In 1865 this description was finally published by Dr. Joseph Leidy, who formally named the species *Astrodon johnstoni,* making it the first sauropod dinosaur described from North America and the first dinosaur described from Maryland. Leidy's description interested Yale University professor O.C. Marsh, who renewed exploration at the Arundel Clay iron pits in 1887. Much of this collecting was done by John Bell Hatcher, a geologist with the U.S. Geological Survey and cohort of Marsh. Hatcher quickly discovered what would become the richest Early Cretaceous dinosaur site ever found along the East Coast, and the most famous dinosaur locality ever discovered in Maryland. This site, part of the Arundel Clay unit, extends from Beltsville to Muirkirk in Prince George's County, and has been dubbed "dinosaur alley."

During the winter of 1887-1888, Hatcher collected hundreds of fossils in dinosaur alley, shipping them to Marsh at Yale. Today many of these fossils reside in the Smithsonian Institution. However, Hatcher's bones were disarticulated, and led to confusion over the proper size and taxonomic relationships of *Astrodon*. Johns Hopkins University professor Arthur B. Bibbins made several more significant discoveries in the late 1890s.

The finds of Hatcher and Bibbins did not parlay into success in the 20th century, though. Only a few more recent *Astrodon* discoveries are noteworthy, and regrettably science has learned little about this giant sauropod during the past hundred years. One of the better subsequent discoveries took place in Washington, D.C., when workers building a water filtration plant found a very large *Astrodon* femur in 1942. Much later, in 1989, Robert Eberle unearthed a partial femur near the city of Arbutus, south of Baltimore. Two years later, in May 1991, Arnold Norden and his children, John and Heather, were hunting for dinosaur bones in the Maryland Clay Products pit near Muirkirk when they came across a meter-long fragmentary femur. These isolated discoveries have provided much of the *Astrodon* fossil material discovered during the past century.

Regardless of its rarity, *Astrodon* is still an excellent choice for Maryland's official state dinosaur. For starters, it was the first dinosaur discovered in Maryland, and the first sauropod found in North America. Second, *Astrodon* is far more abundant and is better known than any dinosaur in the Arundel Clay. Because of these reasons, the Maryland General Assembly passed House Bill 1170 and Senate Bill 520 by wide margins, designating *Astrodon* the state dinosaur. Governor Parris N. Glendening signed these bills into law on October 1, 1998. This designation came six years after a similar *Astrodon* bill narrowly failed, in 1992.

The reason for the success of the second bill can be traced to an enthusiastic Maryland dinosaur specialist, Dr. Peter Kranz. After finding his first dinosaur bone in 1989, Kranz has become a popularizer of Maryland dinosaurs, frequently visiting school classrooms to show off his finds. He also lobbied extensively for the passage of the *Astrodon* bill, even visiting the state capitol building with a man dressed in a dinosaur suit! Kranz often leads digs in the Arundel Clay, where visitors can find their own dinosaur fossils. However, for those who prefer museums, *Astrodon* specimens can be seen at the Peabody Museum, National Museum of Natural History (Smithsonian), and the Maryland Science Center in Baltimore. Kranz is even working on designating 22 acres in Prince George's County as a state dinosaur park, a project that has been in the works since 1995.

Although it is fragmentary and isolated, *Astrodon johnstoni* is historically important. This large beast became the first sauropod discovered in North America and the first dinosaur discovered in Maryland when a chemist found two of its teeth in 1858. Since then other isolated *Astrodon* fossils have been found, but the fact remains that this species is poorly understood. Regardless, visitors to Maryland, and one enthusiastic Ph.D., still enjoy this amazing animal.

FOR FURTHER INFORMATION:

Daemmrich, J., 1998. General Assembly Close to Approving Legislation that will make *Astrodon johnstoni* Maryland's Official Dinosaur. Baltimore Sun, March 31, 1998, p. 5B.

Kranz P.M., 1989. Dinosaurs in Maryland. Maryland Geological Survey, Educational Series, No. 6, 34 p.

—, 1996. Notes on the sedimentary iron ores of Maryland and their dinosaurian fauna. in Brezinski, D.K. and Reger, J.P. (eds.), Studies in Maryland Geology: Maryland Geological Survey, Special Publication No. 3, Maryland Geological Survey, pp. 87-115.

Weishampel, D.B., and Young, L., 1996. Dinosaurs of the East Coast. The Johns Hopkins University Press, Baltimore, 275 p.

Maryland Geological Survey, The Maryland State Dinosaur, http://www.mgs.md.gov/esic/fs/fs12.html

Maryland Manual Online, Maryland State Dinosaur, http://www.mdarchives.state.md.us/msa/mdmanual/01glance/html/symbols/dino.html Maryland

NEW JERSEY

HADROSAURUS FOULKII

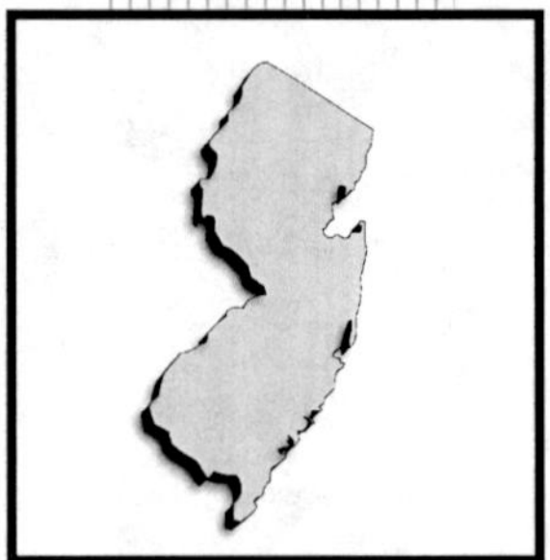

While New Jersey is a small state, it is here that the science of North American paleontology can trace its roots. The beginnings of this now-popular science, the topic of this book and countless others, hark back to the simple discovery of the first relatively-complete dinosaur specimen in the world, a marvelous specimen unearthed in Haddonfield. Later christened *Hadrosaurus foulkii* after its spot of discovery and discoverer, this dinosaur has transformed the study of ancient life like no other creature. It has also been named New Jersey's official state dinosaur.

Hadrosaurus lived during the Cretaceous Period, a time when southern New Jersey was located about 100 feet underwater. The actual coastline was much further east, roughly in the region of what is now Philadelphia. However, after death the bones of *Hadrosaurus* were carried into the shallow sea, and eventually buried under several feet of sediment—luckily for paleontologists.

In life, *Hadrosaurus* was about 25 feet long, about 10 feet tall at the shoulder, and weighed upwards of eight tons. A member of the ornithopod dinosaur lineage, *Hadrosaurus* is an example of what is commonly known as a "duckbill." Most scientists agree that *Hadrosaurus* stood on its hind legs when running, but otherwise was primarily a quadrupedal walker. *Hadrosaurus* roamed the coastal forests and swamps about 80 million years ago, during the Late Cretaceous, feeding on tough-stemmed twigs and leaves. Based on fossil evidence discovered outside of New Jersey, it has been suggested that *Hadrosaurus*, and other duckbills, may have lived and travelled in vast herds. It is known that hadrosaurs laid eggs in nests, and some species cared for their young.

Hadrosaurus foulkii is the best-known hadrosaur from the East Coast, and is known from several fine specimens. However, it is not the only known hadrosaur from this region. Other hadrosaurs include a close relative, *Hadrosaurus minor*, which is known from tail and trunk vertebrae found near Barnsboro, New Jersey; *Hypsibema crassicauda*, discovered in North Carolina; and *Lophorhothon*, known from Alabama. These hadrosaurs share many characteristics, including a long, deep tail used for balance, a padded webbed hand, and a broad, toothless snout. Many of these hadrosaurs are also thought to have been aggressive. *Hadrosaurus foulkii*, for instance, possessed a head perfect for broadside or head-on pushing contests, possibly with rival members of a herd.

Of the above dinosaurs, *Hadrosaurus foulkii* is definitely the most well-known, common, and important. The roots of *Hadrosaurus foulkii* can be traced back to the mid-1800s, when workmen uncovered an amazing specimen in Haddonfield. Around the year 1838, on the farm of John E. Hopkins, an estate known as Birdwood, workers excavating marl for fertilizer hit a cache of petrified bones near a tributary of the Cooper River. After word got out, Hopkins opened his land to tourists and visitors, who carried away the fossils with his permission. Later, some of these specimens were reportedly discovered in southern New Jersey homes as window props and doorstops. Regrettably, Hopkins could not remember to whom he had granted permission to dig up the fossils, and within a short time the cache was depleted.

All Hopkins could recall was that no skull, leg bones, or other large skeletal parts were found, save for a shoulder blade.

Twenty years later, in 1858, William Parker Foulke, a naturalist and member of the Academy of Natural Sciences in Philadelphia, heard about the bones while vacationing in Haddonfield. Based on the descriptions he heard, Foulke reasoned that more bones might be found in Hopkins' marl pit. After talking to the original workmen, Foulke hired a crew to reopen the pit that autumn. He was rewarded when the crew discovered a 75-million-year-old dinosaur skeleton approximately 10 feet beneath the surface. He described this dinosaur as "larger than an elephant, with structural features of both a lizard and a bird."

Recognizing the importance of his specimen, Foulke summoned Dr. Joseph Leidy to the site. Leidy, a university professor and curator of the Academy of Natural Sciences, made the first sketches of the dinosaur based on a careful study of the bones, and later formally presented the skeleton to the scientific community. In an 1858 paper, Leidy described the specimen as, "(consisting) of 28 vertebrae, mostly with their processes lost; humerus, radius, and an ulna, complete; an ilium and a supposed pubic bone, imperfect; a femur and tibia, complete; a fibula, with one end lost; two metatarsal bones and a phalanx, complete; two small fragments of jaws, and nine teeth." Also in this paper, Leidy interpreted the remains to represent a bipedal, amphibious dinosaur, and reconstructed the animal as kangaroo-like. He officially named it *Hadrosaurus foulkii,* in honor of its discoverer.

Ten years later, in 1868, the Haddonfield *Hadrosaurus* became the first dinosaur skeleton to be mounted and put on exhibit, when it was displayed at the Academy of Natural Sciences in Philadelphia. The mount, created by sculptor Benjamin Waterhouse Hawkins, depicted *Hadrosaurus* in what was then thought to be a realistic pose. Since the head and several other skeletal parts were missing, these were approximated with plaster reconstructions. For the first time in history, members of the public could walk up close to a prehistoric creature, making this exhibit extremely popular. During the first year of the skeleton's exhibition, visitation at the Academy more than doubled. Attendance continued to rise until the late 1870s, when other museums around the world began to copy the success of the Academy, and started to display dinosaurs of their own. Soon afterward, the Academy sold duplicate casts to Princeton University and the Smithsonian Institution. One cast was even sold to the Royal Scottish Museum in Edinburgh in 1879, becoming the first dinosaur skeleton to be exhibited in Europe.

Despite the popularity and importance of *Hadrosaurus*, the Haddonfield site on which it was discovered turned to ruin. For many years it was used as an unofficial dumping ground for household wastes, and was quickly forgotten. However, the site was resurrected in 1984, when a local Eagle Scout, 13-year-old Christopher Brees, read about the Haddonfield discovery. When he couldn't find anybody who knew where the site was located, he decided to set out and find it himself. His hard work paid off when he discovered the locality, which he later cleaned and marked, using grants from the Academy and Junior Women's Club of Haddonfield to fund his work. Brees also sculpted a plaque of bronze commemorating the site, and placed it on a boulder nearby.

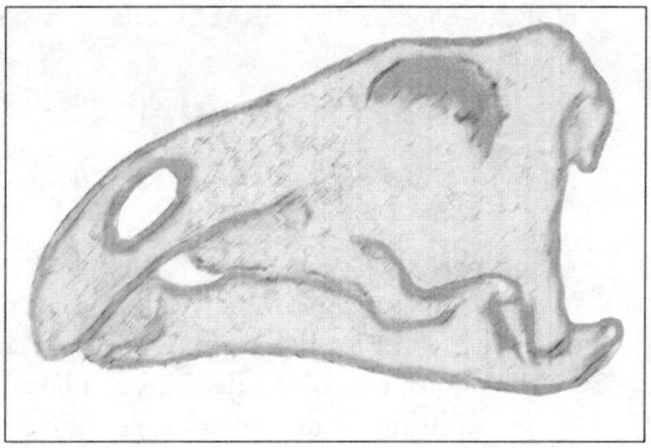

Learning about Brees' work, fourth grade teacher Joyce Berry launched a project aimed at teaching her students about the dinosaurs of New Jersey, along with the function and process of government. Berry guided her students at Strawbridge Elementary School in Haddon Township through the process of proposing to the New Jersey legislature that *Hadrosaurus foulkii* be named the official state dinosaur. Two New Jersey representatives, John Rocco and Thomas Shusted, sponsored the bill, which was later passed and signed into law by Governor James Florio on June 13, 1991. Two years later historian John W. Bond submitted a proposal to the U.S. National Park Service, asking that the Haddonfield site be declared a National Historic Landmark. His dream became a reality in October, 1994.

Because of the work of Brees, Berry, and Bond, tourists can visit the original Haddonfield site today. *Hadrosaurus* reconstructions can also be viewed at the New Jersey State Museum in Trenton and Academy of Natural Sciences in Philadelphia, where some of Foulke's original bones are also on display.

Because of the discovery of the Haddonfield *Hadrosaurus*, the science of North American paleontology was born. Ten years after the discovery, the specimen was cast and mounted, giving the public their first glimpse of what dinosaurs looked like. For many, this specimen proved dinosaurs were real, not creatures of myth. For science, it launched new theories on the posture, size, and diversity of dinosaurs. This discovery changed paleontology from a posh, quaint, Victorian gentleman's hobby to a mainstream scientific activity, and introduced the study of fossils to the American public. For the first time in American history museums became popular places of interest, and commercial fossil hunting became big business. All of this can be traced to a simple discovery, the unearthing of the official state dinosaur of New Jersey, *Hadrosaurus foulkii.*

FOR FURTHER INFORMATION:

Gallagher, W.B., 1997. When Dinosaurs Roamed New Jersey. Rutgers University Press, Camden.

Kurth, S., Fossils of New Jersey, http://home.earthlink.net/~skurth/

Leidy, J., 1858. Remarks Concerning *Hadrosaurus foulkii.* Academy of Natural Sciences of Philadelphia Proceedings, p. 255-257.

—, 1859. Account of the Remains of a Fossil Extinct Reptile Recently Discovered at Haddonfield, New Jersey. Academy of Natural Sciences of Philadelphia Proceedings, p. 1-16.

Levins, H., *Hadrosaurus*, http://www.levins.com/dinosaur.html

Osborn, H.F., 1919. Biographical Memoir of Joseph Leidy, 1823-1891. National Academy of Sciences Biographical Memoirs vol. 9, p. 339-396.

Warren, L., 1998. Joseph Leidy: The Man Who Knew Everything. Yale University Press, New Haven.

Weishampel, D.B., and Young, L., 1996. Dinosaurs of the East Coast. The Johns Hopkins University Press, Baltimore, 275 p.

TEXAS

Pleurocoelus

Everything is bigger in Texas, which makes it no surprise that Texas' official state dinosaur is one of the largest known from the American south. Although its early discovery is unclear, its relationship with preserved footprints is unsettled, and debate still surrounds its proper taxonomic name, *Pleurocoelus* is a fitting state dinosaur for the Lone Star State.

The story of *Pleurocoelus* begins about 110 million years ago, in the Early Cretaceous. At this time much of Texas was a muddy tidal flat, located at the edge of a shallow sea that continually advanced and retreated across central North America for millions of years. These wide flats, covered by bushes, shrubs, and trees, were the perfect spot for giant sauropods to browse. *Pleurocoelus* was the largest of these sauropods.

Pleurocoelus was a brachiosaurid, closely related to *Brachiosaurus*. It was a large, sturdy, quadrupedal herbivore, whose front limbs were as long as or longer than its hind limbs. *Pleurocoelus* had a small, narrow skull, with teeth similar to those of *Camarasaurus*, but thinner. Its limbs also resembled *Camarasaurus*', but *Pleurocoelus*' toes were more slender than those of most sauropods. Based on footprint evidence, the fourth toe of the hind foot may have had a small claw, an uncommon feature in sauropods, while the front foot was likely without claws. The claw was apparently small, and was likely buried in the tissue of the sole of the foot. This anatomy makes the function of the claw unknown, although some think it may have been a defensive weapon.

Like other brachiosaurids, *Pleurocoelus* was built something like a large giraffe—an analogy brought about by its longer front legs, high shoulders, sloping back, and vertical neck. This neck likely meant that *Pleurocoelus* was a high browser, able to feed on the leaves of trees too high for other herbivores to reach. *Pleurocoelus* was a relatively small brachiosaurid, although some believe that most known skeletal fossils represent young individuals. While some aspects of dinosaur bone histology are still poorly understood, it is thought by many that the bones of *Pleurocoelus* indicate fast, uninterrupted growth from the juvenile to the adult stage. This may mean that *Pleurocoelus* was not ectothermic (cold-blooded), a notion supported also by the vaulted nares (nostrils) on the top of its skull, possibly a vent for excess body heat.

Although complete skeletons are unknown, *Pleurocoelus* individuals were anywhere from 8-20 meters long and weighed from 20-45 tons. The name *Pleurocoelus* means "hollow side," which refers to the pleurocoels, or air spaces, in its vertebrae.

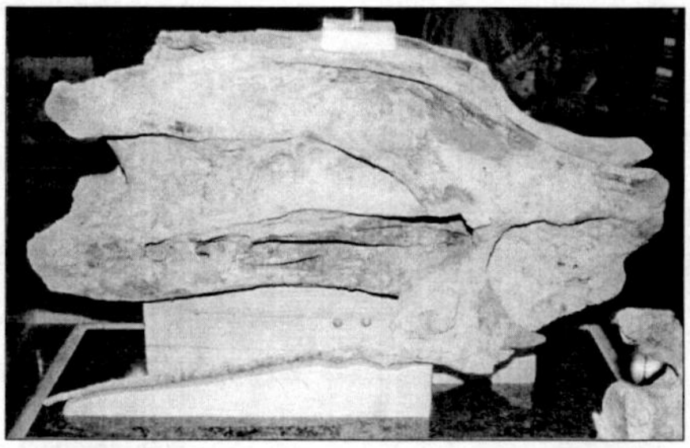

Bones of *Pleurocoelus* are found in the Early Cretaceous Glen Rose Limestone of central Texas, the same unit in which tracks of the ichnogenus *Brontopodus* are found. These *Brontopodus* tracks are most definitely those of sauropods, and because bones of *Pleurocoelus* are often discovered close to these tracks, many scientists have suggested that *Pleurocoelus* was the *Brontopodus* trackmaker. While this cannot be conclusively proven, recently discovered pedal (foot) material attributed to *Pleurocoelus* matches closely with the *Brontopodus* tracks, lending credence to this hypothesis.

Brontopodus tracks are commonly discovered in the Glen Rose Limestone, and tend to be distributed in zones at the top of strata in this formation. One of the most popular Glen Rose Limestone track sites is also one of the most fiercely debated. Located near the city of Glen Rose southwest of Fort Worth, Dinosaur Valley State Park along the Paluxy River has a long history, and the debate over the evidence it offers still continues.

Although it is unknown who discovered the first *Pleurocoelus* bones in Texas, it is well-known who discovered the Paluxy River site. During the late 1800s, local moonshiners and fishermen commonly spoke of "tracks in stone" discovered near the Paluxy River. One of these prints was even built into the bandstand at the Somervell County courthouse in Glen Rose. However, paleontologists didn't learn about these tracks until the early part of the 20th century, when American Museum of Natural History paleontologist Roland Bird noticed a fake fossil print for sale at a roadside stand in Texas in 1938. He knew the print was a fake, but he inquired around and was told by locals that tracks were commonly reported from near Glen Rose. He made his way to the Paluxy River, where he discovered a whole slew of dinosaur footprints. Among these tracks were those of theropods and at least a dozen sauropod prints, now described as *Brontopodus*.

Bird contacted a crew from the Works Progress Administration, ordered the crew to divert the river, and chipped out huge sections of the trackway that were later sent to New York to be reassembled. Since Bird's initial discovery, at least 1,000 tracks have been found near the Paluxy River. As would be expected, these discoveries made the Glen Rose site popular, and several locals attempted to buy the land in the early 1960s, hoping to find more prints. However, it was the Texas Parks and Wildlife Department which bought the property in 1969, and opened it to the public the next year. Today the area is known as Dinosaur Valley State Park, and it is one of the most popular tourist destinations in central Texas.

Many of the visitors to the park are drawn because of an intriguing hypothesis first popularized by Bird. After uncovering a dozen parallel *Brontopodus* trackways along the Paluxy, Bird discovered three theropod trackways which followed the same trend as the sauropod prints. Upon further examination, Bird noticed the tracks of a single theropod and a single sauropod, preserved as if they were walking near each other. Furthermore, one of the theropod tracks in the sequence was reported as missing. To Bird, this was unquestionably the site of a predator-prey chase, with the missing theropod print possibly indicating that the predator had attacked the sauropod by latching onto its left rear flank, before it resumed normal progression. However, many paleontologists disagree. One of these trackers, Martin Lockley, points to the fact that there is little evidence the animals changed speed or direction as they would during a pursuit. Also, Lockley said, there is no way of knowing if the theropod tracks were made at the same time as the sauropod's. Instead, he wrote, twelve sauropods, possibly *Pleurocoelus*, probably as a herd, were followed somewhat later by three theropods that may or may not have been stalking, but were certainly not attacking. Regardless, the site draws thousands of visitors each year, largely due to this supposed ancient chase.

In addition to Dinosaur Valley State Park, *Brontopodus* and other sauropod tracks can be seen at several other locations in Texas, among them one in Bandera County, where Bird reported a sequence of sauropod tracks consisting only of forefoot prints in a straight line. Based on this evidence, Bird proposed in a 1944 paper that the animal was afloat, and only dragged its forefeet as it floated. This interpretation has been challenged, but many do believe it accurate. Just recently another manus-only sauropod trackway found in Korea has been interpreted in the same manner. Other tracks have been discovered near the West Verde Creek in Medina County, near Sidney in Comanche County, along the Blanco River in Blanco County, and along the San Gabriel River in Williamson County. Tracks and/or *Pleurocoelus* material can also be seen at the American Museum of Natural History in New York, the Texas Memorial Museum at the University of Texas in Austin, and the Fort Worth Museum of Science and History.

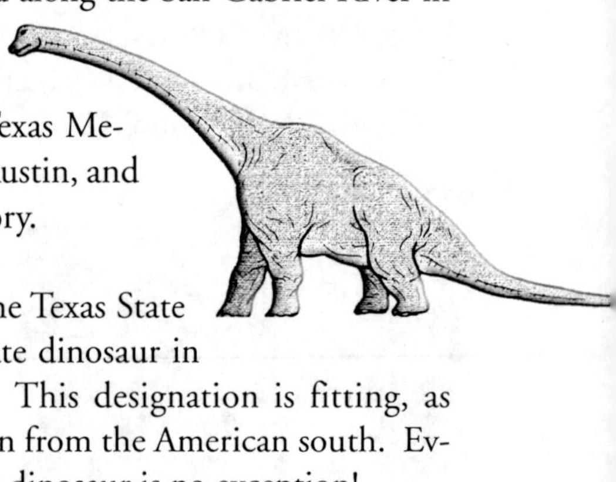

Regardless of the abundance of tracks, the Texas State Legislature named *Pleurocoelus* the official state dinosaur in 1997, signing Concurrent Resolution #57. This designation is fitting, as *Pleurocoelus* is one of the largest animals known from the American south. Everything is bigger in Texas, after all. The state dinosaur is no exception!

FOR FURTHER INFORMATION:

Bergen, G.R., and Pittman, J.G., 1990. Nearshore Clastic-Carbonate Facies and Dinosaur Trackways in the Glen Rose Formation (Lower Cretaceous) of Texas. Geological Society of America Field Trip #8, Dallas, Texas.

Blair, W. F., 1958. Distributional patterns of vertebrates in the southern United States in relation to past and present environments. in Hubbs, C. L., ed., Zoogeography: American Association for the Advancement of Science. Publication 51, p. 433-568.

Brusatte, S., 2002. Kids' Corner-*Pleurocoelus*. Fossil News: Journal of Avocational Paleontology, v. 8, no. 2, p. 8.

Farlow, J.O., 1992. Of Tracks and the River, in W.H.Shore, ed., Mysteries of Life and the Universe: New Essays from America's Finest Writers on Science, Harcourt Brace Jovanovich, New York, p. 63-84.

—, 1993. The Dinosaurs of Dinosaur Valley State Park. Texas Parks and Wildlife Press, Austin.

Gallup, M. 1989. Functional Morphology of the Hindfoot of the Texas Sauropod *Pleurocoelus* sp. indet. Geological Society of America Special Paper (Regional Studies), no. 238 p. 71-74.

Gillette, D.D., and Lockley, M.G. eds., 1989. Dinosaur Tracks and Traces. Cambridge University Press, Cambridge.

Jacobs, L.J., 1995. Lone Star Dinosaurs. Texas A & M University Press, College Station.

Keesey, T.M., Dinosauricon-*Pleurocoelus*, http://dinosauricon.com/genera/pleurocoelus.html

Lee, Y.N., and Huh, M., 2002. Manus-only sauropod tracks in the Uhangri Formation (Upper Cretaceous), Korea and their paleobiological implications. Journal of Paleontology, Vol. 76, No. 3, pp. 558-564.

Lessem, D., and Glut, D.F., 1993. The Dinosaur Society Dinosaur Encyclopedia. Random House, New York.

Lockley, M.G., and Hunt, A.P, 1995. Dinosaur Tracks and Other Fossil Footprints of the Western United States. Columbia University Press, New York.

Lundelius, E.L., Vertebrate Paleontology of the Balcones Fault Trend, http://www.lib.utexas.edu/geo/BalconesEscarpment/pages41-50.html

Utah Geological Survey, Early Cretaceous Dinosaurs, http://www.ugs.state.ut.us/dinecret.htm

Dinosaur Valley State Park is on the western outskirts of Glen Rose off of US 67; take FM205 north four miles to Park Road 59. For information on the park, call (254) 897-4588.

WYOMING

TRICERATOPS

Like Maryland, Wyoming has designated both a state fossil and a state dinosaur. Unlike Maryland, however, Wyoming's state dinosaur is well-understood, and well-known to the public. In addition this state dinosaur, *Triceratops*, is also the official state fossil of South Dakota. However, unlike with South Dakota, some of the earliest *Triceratops* discoveries can be traced to Wyoming, and it is in this state where the bulk of modern *Triceratops* fossils are found.

Triceratops was a member of the ceratopsian lineage of dinosaurs, and lived in the western United States during the Late Cretaceous. It is possible that *Triceratops* was one of the last dinosaur genera to evolve, as it lived just prior to the time when the dinosaurs became extinct. *Triceratops* specimens are immediately recognized by their large frill and three horns. Although a complete *Triceratops* has never been discovered, over 100 skulls from this genus have been found and described. While it is often hard to tell from bones, the average *Triceratops* individual grew to lengths of at least eight meters (25 feet), and may have weighed upwards of 4.5 metric tonnes (5 tons).

The most distinctive part of *Triceratops*' anatomy was, and is, undoubtedly its skull. This skull, which grew to lengths of eight feet, was characterized by the aforementioned frill and three horns. Two of these horns were positioned behind the eyes of *Triceratops*, and are known as brow horns. The third horn, known as the nasal horn, was located above the nose. While the nasal horn was short, the brow horns often grew as long as three feet. All three horns were anchored by a bony horn core, an outgrowth of the skull bones, and covered in a tough mesh of keratin. Also located on the skull was a large frill, composed of the parietal and squamosal bones, which grew to lengths of six or seven feet. The frill of *Triceratops* is one of the only known ceratopsian frills to lack parietal fenestrae (frill holes).

The jaws of *Triceratops* occupied the anterior portion of the skull. These jaws were powerful and equipped with teeth that looked much like elongated blades. These teeth were also strong, and likely used to grind up tougher plant material, such as cycads and palm fronds. The strong jaws of *Triceratops* were powered by large muscles anchored to the frill, a possible reason for the large size of the frill. It is also thought that the frill may have been a display organ, used to attract mates or intimidate other members of the species.

The body of *Triceratops* was strong and sturdy, and was carried on four legs. There is debate over the proper orientation of these legs, with some scientists believing that they were carried directly underneath the body, mammal-style, while others suggest a more sprawling, lizard-like stance. Regardless, *Triceratops* was likely rather slow, but could charge if necessary, much like a modern rhinoceros. This ability to charge may have been a defense mechanism. The three horns were probably useful for deterring potential predators. One skull, discovered in South Dakota, shows probable *Tyrannosaurus* bite marks, giving support to this theory. However, some paleontologists have hypothesized that the horns were display structures only. The most likely answer is that they served both functions.

Since the first *Triceratops* discovery, some 100 skulls of this genus have turned up, many of them in Wyoming. The first record of *Triceratops* can be traced back to around 1872, when E.D. Cope described a few vertebrae and a partial sacrum from an animal he called *Agathaumus*. These remains came from near Black Butte, Wyoming, and intrigued Cope, who sent collectors out into the field to discover more specimens. Later, in 1887, a pair of large fossil horn cores was sent to O.C. Marsh, Cope's vehement rival, by government geologist Whitman Cross, who discovered the specimens near Denver, Colorado. Marsh incorrectly concluded that these cores represented an extinct buffalo, describing them as *Bison alticornis* in 1888. Later that year, however, his views would change, as one of his collectors, John Bell Hatcher, came across a large horn core while prospecting near Lusk, Wyoming. He wrote to Marsh describing the discovery, and later sent the professor the fossils. These cores finally convinced Marsh that what he called *Bison alticornis* was not a buffalo, but an extinct dinosaur. In 1889 Marsh officially named *Triceratops* based on these remains.

Hatcher's discovery turned out to be the first of many ceratopsian finds in the Lance Formation sediments of Niobrara County (then known as Converse County), Wyoming. The Lance Formation is interesting because it represents an ancient floodplain in a warm, temperate seasonal climate. This unit, which outcrops over much of Niobrara County, is up to 750 meters thick in places, and covers 1.5 million years of geological time. Countless fossils have been found here, including bones of *Tyrannosaurus, Ankylosaurus, Dromaeosaurus, Pachycephalosaurus*, and, of course, *Triceratops.*

After his discovery of the Lusk horn cores, Hatcher was sent back into Wyoming by Marsh. During the next three years, from 1889-1892, he collected some 40 *Triceratops* skulls, including those

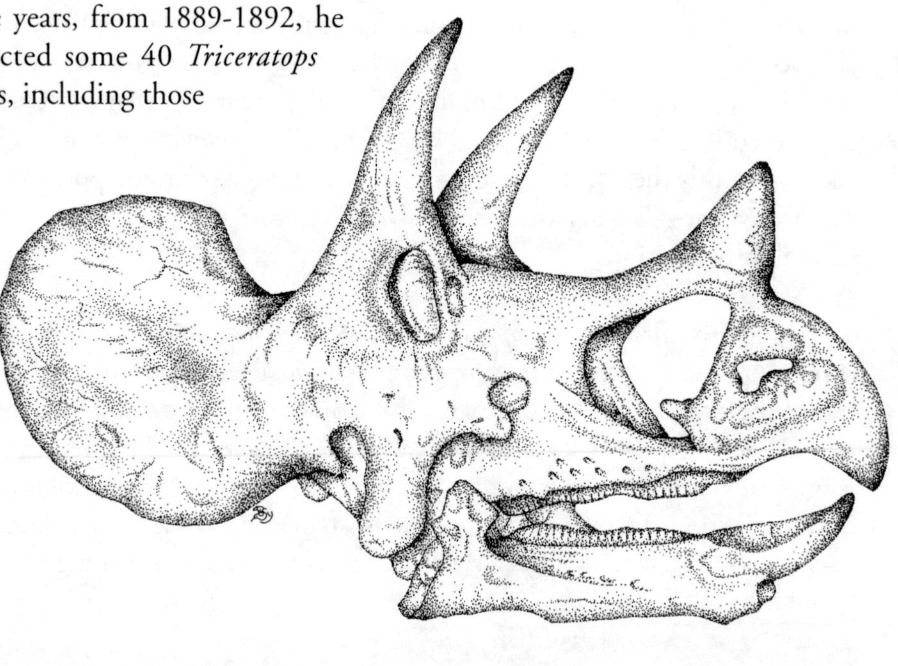

referred to a variety of dubious species. One of the specimens collected by Hatcher is currently on display at the Smithsonian's National Museum of Natural History in Washington, D.C. This specimen, catalogued as USNM 2580, is composed of bones from seven individuals. When it was mounted in 1905, this *Triceratops* became the first specimen of this genus to be displayed. In charge of the preparation and mounting of these bones were Charles Gilmore and Norman H. Boss. While their work was admirable for the turn of the last century, today it has been recognized that several aspects of their restoration were incorrect. The mount was also starting to disintegrate due to the lack of environmental controls and the effects of the continuous hordes of Smithsonian tourists. In October, 1998 the specimen was dismantled, and for the better part of two years it was reassembled. In May, 2001 the new and improved mount was unveiled to the public. The Smithsonian sponsored a nickname contest, which resulted in the specimen being named "Hatcher" in honor of its collector.

Lost in the glorified mounting of the Hatcher specimen was the fact that the taxonomy of *Triceratops* was becoming heavily flawed. All in all, sixteen species of this genus have been named, most of which came from Niobrara County. It is very unlikely that sixteen species of the same genus, of the same general size, lived at the same general time in the same general area—something is most definitely wrong. Finally, some of the mess was cleared up in a 1986 paper written by John Ostrom and Peter Wellenhofer. In this paper, the two scientists sank fifteen of the *Triceratops* species, and recognized only one: *Triceratops horridus,* the species first named by Marsh in 1889. However, in her 1990 Ph.D. thesis, Cathy Forster recognized two species of *Triceratops: T. horridus* and *T. prorsus.* Regardless of whether one species or two ultimately stand, it is refreshing to see that most of the quagmire over the proper names and species of *Triceratops* has been resolved.

Triceratops was officially named the Wyoming state dinosaur in 1994, after legislation providing for a contest and election by elementary school children was carried out. *Triceratops* was chosen by the children over three other candidates. Luckily, for these children, *Triceratops* remains and mounts can be seen at several museums in the state, including the University of Wyoming Geological Museum, the Wyoming Dinosaur Center in Thermopolis, and Western Wyoming Community College.

Wyoming is one of two states that recognize both an official state fossil and an official state dinosaur. However, unlike the enigmatic *Astrodon* of the Old Line State, Wyoming's state dinosaur is both well-known and extremely popular with the residents of the state, as evidenced by the hordes of elementary students who voted for this Cretaceous ceratopsian as an official symbol. There are few things more popular with young children than dinosaurs, and of these dinosaurs, there are few more popular than *Triceratops*.

FOR FURTHER INFORMATION:

Bigelow, P., Hell Creek Faunal Facies, http://www.scn.org/~bh162/hellcreek2.html

Dodson, P., 1996. The Horned Dinosaurs. Princeton University Press, Princeton, New Jersey, 346 p.

Forster, C.A., 1990. The Cranial Morphology and Systematics of *Triceratops* with a Preliminary Analysis of Ceratopsian Phylogeny. Pennsylvania: University of Pennsylvania Department of Geology, unpublished Ph.D. thesis, 227 p.

Hatcher, J.B., Marsh, O.C., and Lull, R.S., 1907. The Ceratopsia. United States Geological Survey Monograph 49: 1-300.

Holmes, T. and Holmes, L., 2001. Horned Dinosaurs: The Ceratopsians. Enslow Publishers, Berkeley Heights, New Jersey, 128 p.

Lehman, T.M., 1990. The Ceratopsian Subfamily Chasmosaurinae: Sexual Dimorphism and Systematics. in: Dinosaur Systematics-Approaches and Perspectives, Carpenter, K., and Currie, P.J., eds, Cambridge University Press, Cambridge, pp. 211-229.

Marsh, O.C., 1888. A New Family of Horned Dinosaurs from the Cretaceous. American Journal of Science, ser. 3, 36: 477-478.

—, 1889a. Notice of New American Dinosauria. American Journal of Science, ser. 3, 37: 331-336.

—, 1889b. Notice of Gigantic Horned Dinosauria from the Cretaceous. American Journal of Science, ser. 3, 38: 173-175.

Ostrom, J.H., and Wellenhofer, P., 1986. The Munich Specimen of *Triceratops* with a Revision of the Genus. Zitteliana 14: 11-158.

—, 1990. *Triceratops*: An Example of Flawed Systematics. in: Dinosaur Systematics-Approaches and Perspectives, Carpenter, K., and Currie, P.J., eds, Cambridge University Press, Cambridge, pp. 245-254.

University of Wyoming, *Triceratops*, http://www.uwyo.edu/geomuseum/tricer.htm

State Stones and Gems

FLORIDA

AGATIZED CORAL

The state of Florida preserves a host of interesting fossils. While most fossils are found deeply embedded in rock units, Florida fossils are most commonly found weathered away from rock. Many times, paleontologists have to use careful preparation methods just to free fossils from their hard, tough matrix, but Florida fossil hunters only have to wipe away the dust from their specimens. And, when many fossil hunters must endure long, hard treks to fossil sites, most Florida rock hounds only have to don some scuba gear and they are set.

Few would argue with the notion that fossil hunting in Florida is much different from collecting in most other states. Because of Florida's abundance of streams and rivers, flowing water does the job of liberating fossils from the rock. All fossil hunters have to do is know where to look and how to swim. The majority of these river fossils represent those of Tertiary specimens, including saber-tooth cats, giant ground sloths, and glyptodonts. However, as yet, Florida does not have an official state fossil. The Sunshine State does recognize a fossil as its state stone, though.

While the aforementioned Tertiary river specimens are the most commonly collected Florida fossils, the state gem does not fit the above mold. Instead, the state gem, agatized coral, is not commonly found in rivers, and was not a large, land mammal. This animal lived in the vast Eocene seas that covered much of Florida about 35-40 million years ago, long before the ground sloths and saber-tooths left the remains that today are commonly found by divers.

Agatized coral is not phylogenetically any different than modern coral. Exactly what is a coral? Corals are invertebrate animals, and members of the phylum Coelenterata. Like all other members of the phylum, coral polyps are radially symmetrical and have an internal space for digestion, called the gastrovascular cavity. This cavity opens into a mouth, which is surrounded by tentacles that aid in food capture.

Corals come in two major forms: solitary and colonial. The agatized coral commonly found in Florida was a member of the colonial stock. What this means is that individual coral animals, known as polyps, gathered into large colonies attached to hard surfaces. When alive, polyps combine their own carbon dioxide with lime in the seawater to form a limestone-like hard exoskeleton. This exoskeleton, which composes the familiar structure of coral reefs, serves a dual purpose: protection and support. In addition to simply anchoring the soft body of the polyp, it also serves as a chamber into which the coral animal can retreat when threatened by predators.

Corals and coral reefs are very familiar, but what about agatized coral? Agatized coral occurs when silica in the ocean water hardens, replacing the limey exoskeleton of corals with a form of quartz mineral known as chalcedony. This chalcedony, which is extremely durable, comes in a variety of colors, many of which are discussed in the section on West Virginia's state gem, the agatized coral *Lithostrotionella*.

It takes agatized coral an exhaustingly long time to fossilize, often 20-30 million years. Therefore, it can form only under the most ideal conditions. Today, some of the best agatized corals can be found in Florida's Coastal Plain, specifically in the Tampa Bay area, along the Econfina River, and in the Withlacoochee/Suwannee River beds.

Fossils of these corals are commonly exposed along rivers, in road cuts, and in limestone quarries. These limestone quarries are remnants of large Eocene seas, which had the perfect temperature and chemical composition to form vast limestone deposits. Today, most of these limestones harbor fossils of a diverse variety of marine animals, including several species of coral.

While this coral is very abundant in Florida, it was not the State Legislature's choice for official state gem. This honor went to the moonstone, a rock recognized after the successful Apollo space missions in the 1960s and 1970s. In 1979, however, the Florida State Gem and Lapidary Society became upset that Florida's official state gem wasn't even found in the state, so they recommended that the Legislature designate a state stone. The Society promoted agatized coral, which was officially adopted as the state stone later that year.

While Florida has a vast fossil record consisting of everything from large land mammals to prehistoric sharks to small marine invertebrates, the Sunshine State has not designated an official state fossil. However, naming a fossil as Florida's state stone makes up for this vacuum.

FOR FURTHER INFORMATION:

Anderson, R., 1988. Guide to Florida Rocks, Minerals, Fossils. Winner Enterprises.

Randazzo, A., 1997. The Geology of Florida. University Press of Florida.

Renz, M., 1999. Fossiling in Florida. University Press of Florida.

Division of Historical Resources, The Florida State Stone, http://dhr.dos.state.fl.us/symbols/stone.html

MICHIGAN

PETOSKEY STONE

Unlike most states, Michigan does not have an official state fossil. However, like a handful of others, the Great Lake State has designated a fossil as its official state stone. This stone, commonly found in the northern portion of the Lower Peninsula, is one of the most popular geological specimens with local rockhounds. It is also commonly seen in souvenir stores, as worry stones, or embedded in jewelry. What exactly is this famous rock? The Petoskey Stone, of course!

Petoskey Stones are actually small chunks of Devonian coral reefs, broken off the bedrock by glacial action. The coral in question belonged to the species *Hexagonaria percarinata*, one of the most easily recognized Devonian tabulate corals. During the time *Hexagonaria* lived, about 350 million years ago, Michigan was positioned near the equator, and covered by a shallow, warm sea. This large sea was an ideal habitat for marine life, and sported vast coral reefs. These reefs supported a wide array of life forms, including clams, cephalopods, crinoids, trilobites, and fish. While fossils of all of these organisms have been found in Michigan's Devonian sediments, it is the fossils of *Hexagonaria* that are most sought after, and most commonly found.

Like other corals, *Hexagonaria* consisted of a soft, living part, called the polyp, and a hard, outer shell. At the center of the polyp was the mouth, which was surrounded by tentacles. These tentacles were used to sweep food and water into the coral individual's rudimentary gut. Most of this food likely consisted of plankton, a common food for modern corals and one known from Michigan's Devonian fossil record. Individual *Hexagonaria* corals, which are recognized on the basis of their hexagonal shape, would gather into large reefs, which were often composed of millions of single coral animals. The reefs were basically formed by the secretion of a hard, limey shell by each individual coral, a structure used as a skeletal base to support the polyp. Over time, calcite, silica, and other minerals replaced the original elements in each coral, fossilizing parts of the reefs. Today, small parts of these coral reefs are commonly found as Petoskey Stones.

Petoskey Stones are well-known from the Middle Devonian Gravel Point Formation, and were first discovered near the town of Petoskey. They are commonly found along beaches in the Little Traverse Bay, on the Lake Michigan side of the state, although specimens have been found along the shores of Lake Huron. Most specimens are found in a string of northern Michigan counties, including Charlevoix, Emmet, Cheboygan, Presque Isle, and Alpena counties. However, often Petoskey Stones are found quite far from where they were originally deposited in the Devonian. The reason for this lies in glacial action. During the last great Ice Age, glaciers overran bedrock units, literally plucking off parts of the rock. These small rocks, often the remains of *Hexagonaria*, would become trapped in the glacier, and would be deposited in remote areas after the glaciers melted. This explains why Petoskey Stones are often found miles from Devonian outcrops, and why isolated specimens often turn up in fields.

Because of the action of glaciers and meltwater rivers, Petoskey Stones are usually rounded and smooth. Common specimens range from the size of a quarter to that of a potato, although much larger Petoskey Stones have been found. Good specimens can be found along beaches, in road cuts, and embedded in fields throughout the northern part of the Lower Peninsula, although one of the best places to hunt for these strange stones is in Fisherman's Island State Park near Charlevoix. This park, which lies along Lake Michigan, is a well-known collecting site. Other great spots for Petoskey Stone collecting are along the south end of the beach at Petoskey State Park and at Magnus City Park.

While the Petoskey Stones found in outcrops and along beaches are often rounded and smoothed, it is additional polishing by collectors and rockhounds that gives these stones their classic look, and make them a popular souvenir for any visitor to Michigan. Since Petoskey Stones are primarily calcite, they are very easy to polish without ruining the specimen. While not all specimens are suitable for polishing, the best Petoskey Stones are usually polished by hand. This polishing is quite easy, and can be accomplished using minimal materials, among them a thick towel, sandpaper, polishing powder, water, and a piece of corduroy or velvet.

To begin polishing, start by filing the stone with the sandpaper until you obtain a suitable size and shape. Then, simply dampen the stone with water and rub the specimen with 220 grit sandpaper with a steady, rotating motion. After rubbing, rinse the stone again and dry it. Examine for scratch marks, which should be removed by additional rubbing. Repeat the process using 400 grit sandpaper, and complete with 600 grit paper. After completing the sanding, rinse and dry the stone. Using a piece of corduroy or velvet, rub the stone with a polishing powder or polishing compound. A short, rotating rubbing does the trick. When the polishing is complete, wipe off the stone with a clean, dry cloth. If the steps are followed correctly, the stone should be perfect for your collection! If not, you can simply buy a polished stone at most northern Michigan souvenir stores.

It is unclear who discovered the first Petoskey Stone, although it is likely that Native Americans came across them many centuries ago. In fact, the city of Petoskey itself is deeply-rooted in Native American legend. According to an

Ottawa tale, a descendant of French nobility named Antoine Carre visited what is now the Petoskey area and became a fur trader with the Astor Company. He married a native princess, and later became the tribe's chief. In the spring of 1787, a son was born to Carre and his wife. When the chief saw the rising sun's brilliant rays reflected on his son's face, he proclaimed, "his name shall be Pe-tos-e-gay." This name means "rising sun," and is the etymology for the city of Petoskey. Later in life, Pe-tos-e-gay grew up to become a wealthy fur trader. In the summer of 1873, shortly before Pe-tos-e-gay's death, a city was formed along the bay at Bear Creek. It was named Petoskey in his honor.

This legend has little to do with the Petoskey Stone itself, but is interesting because the polished Petoskey Stone reminds some of the rays of the rising sun! The first true scientific study of this famous stone was published in 1969 by Dr. Edwin C. Stumm, who named the coral that makes up the stone *Hexagonaria percarinata.* However, four years before his paper, on June 28, 1965, Governor George Romney signed Public Act No. 89, which designated the Petoskey Stone Michigan's official state stone. This legislation was the first such act in the United States to declare a fossil a state stone, a procedure several other states have since followed. It was fitting that Miss Ella Jane Petoskey, the only living grandchild of Pe-tos-e-gay, was on hand for the designation!

Interesting Petoskey Stone discoveries continue today. In the late 1990s amateur fossil hunter Chuck Schnake discovered a massive Petoskey Stone at Sleeping Bear Dunes National Lakeshore. This large stone, which measures 40 inches long by 20 inches deep, may weigh up to 2,000 pounds. Schnake believes it is the biggest Petoskey Stone ever unearthed, and is in the process of having it displayed at the state capitol.

While Michigan does not have an official state fossil, its state stone is likely the most important and popular fossil to have ever been discovered in the state. Known as the Petoskey Stone, this state stone actually belongs to the Devonian coral species *Hexagonaria percarinata*. Commonly unearthed by rockhounds, polished by jewelry makers, and purchased in souvenir stores, this stone also gives scientists clues about the ancient environment of Michigan. Perhaps no other state stone is as popular, or as important.

FOR FURTHER INFORMATION:

Brusatte, S., 2001a. Kids' Corner-Petoskey Stone. Fossil News: Journal of Avocational Paleontology, vol. 7, no. 2, p. 9.

—, 2001b. The Petoskey Stone. West Virginia Fossil Club Newsletter, May 2001, p. 4.

Fochtman, B., Petoskey Stones, http://www.boynecountry.com/ptstone.html

Kelley, R.W. Guide to Michigan Fossils. Michigan Department of Conservation.

Raymer, M. Rockhound Unearths Massive Petoskey Stone at Sleeping Bear. Record-Eagle, Date Unknown.

Teske, E., Petoskey Stones, http://www.geo.msu.edu/geo333/petoskystone.html

Michigan Department of Environmental Quality, The Petoskey Stone-The Michigan State Stone, http://www.deq.state.mi.us/gsd/petoskey.html

Petoskey Regional Chamber of Commerce, Petoskey Stones, http://www.petoskey.com/pstone.html

MISSISSIPPI

PETRIFIED WOOD

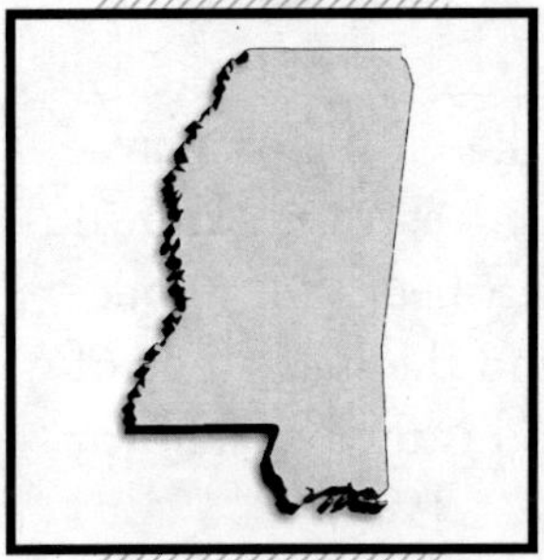

Mississippi does recognize an official state fossil, the prehistoric whale. In addition, the Jewel of the South also has designated a fossil as its state stone. However, like the state fossil, this designation is also ambiguous. The State Legislature simply recognized "petrified wood," a title that covers a plethora of different fossils found throughout Mississippi's Coastal Plain.

If you recall from the section on Louisiana's official state fossil, what we call petrified wood is nothing more than fossilized tree stems, trunks, and branches, and often consists of a variety of minerals. These minerals, which include silica, sulfates, sulfides, oxides, silicates, carbonates, and phosphates, permeate the wood after it dies and is buried, and over time turn it to stone. These petrified specimens are often found in two types of sediments: in accumulations of volcanic ash and breccia or in sediments deposited by rivers and streams. The petrified wood found in both Louisiana and Mississippi is known from the latter type of deposit.

In reality, much of the petrified wood found in Mississippi and Louisiana is very closely related, both phylogenetically and geographically speaking, having formed in nearly the same way. After the tree died and was buried, dissolved silica within the groundwater attached itself to the lignin and cellulose of the wood. Over time, a layer of silicic acid formed on the exposed woody tissues. This acid later dehydrated into silica gel, which over time eventually filled the wood completely. The gel absorbed the water in the wood, desiccating it and converting the silica gel into amorphous silica or opal. Over the next several million years, this opal dehydrated and turned into microcrystalline quartz, or chert. During this process, various minerals became incorporated into the silica gel, causing it to change colors. As a result, the petrified wood found in Mississippi today can range in color from near black to blue to red.

Several varieties of petrified wood are known from the Gulf Coast area, including a nondescript type known as "silicified wood," a denser type known as "massive silicified wood," and a third kind known as "petrified palm wood." This petrified palm wood is the state fossil of Louisiana and the state stone of Texas. Belonging to the form genus *Palmoxylon*, this wood is the best known from the tri-state area, and is very popular with collectors.

Petrified palm wood is basically a group of fossil woods that contain prominent rod-like structures within the regular grain of the silicified wood. These structures, although altered, are the remnants of sclerenchyma bundles that once comprised part of the woody tissues that gave the stem its vertical strength.

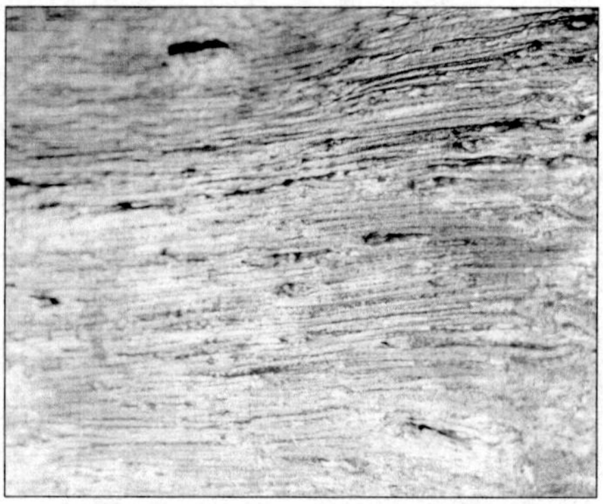

The petrified palm wood found in the area comprised a goodly number of species. Three distinct species, *Palmoxylon remotum, Palmoxylon mississippense*, and *Palmoxylon ovatum*, are known from Mississippi. Several others are known from Texas and Louisiana. Individual trees of these species grew in what is now Mississippi some 30 million years ago, during the Oligocene Epoch. During the time the Gulf of Mexico's shoreline was much farther north, which explains why petrified wood is found in more northern areas of both Louisiana and Mississippi. Although it is difficult to deduce much about the plant's appearance and lifestyle from isolated pieces of wood, comparisons with modern palm trees are used to help reconstruct these

Oligocene palms. Therefore, it is assumed that Oligocene palms had no branches and all of their leaves were clustered near the top.

Many of the best petrified wood specimens in Mississippi come from the Mississippi Petrified Forest. Located at 124 Forest Park Road in Flora, this fossil site has been known since the mid-1800s, but only within the last three decades has it been developed and opened to the public. In 1966 the area was officially declared a Registered National Natural Landmark by the National Park Service. Today, this area, the only petrified forest in the eastern United States, is privately owned and operated. The park brags that over "36 million years of geological history are preserved in stone." It is easy for visitors to get lost in the park's nature trails, earth science museum, and outstanding collection of petrified wood specimens, some of which are the most spectacular known from the American south.

Ten years after the National Park Service designated the Mississippi Petrified Forest as a landmark, the Mississippi State Legislature officially designated petrified wood as the state stone. Adopted on May 14, 1976, a Senate Concurrent Resolution recognized the importance of this stone, its popularity with collectors, and its discovery at the Mississippi Petrified Forest.

FOR FURTHER INFORMATION:

Blackwell, W. H., 1984. *Palmoxylon* from Bayou Pierre, Copiah/ Claiborne County Line, Southwestern Mississippi. American Journal of Botany, vol. 71, no. 5, Part 2, p. 113.

Blackwell, W. H., Powell, M. J., and Dukes, G. H., 1983. Fossil Wood from Bayou Pierre and White Oak Creek, Southwest-Central Mississippi. Mississippi Geology, vol. vol. 4, no. 2, p. 1-6.

Dockery, D. T. III, 1987. Petrified Palm "Wood" from Thompson Creek, Yazoo County, Mississippi. Mississippi Geology, vol. 8, no. 2, p. 10-11.

Mississippi Petrified Forest, http://www.mspetrifiedforest.com/

TEXAS

PETRIFIED PALM WOOD

It is probably fairly obvious by now that petrified palm wood, known from the Louisiana-Mississippi-Texas area, is a very popular state symbol. Along with being the official state fossil of Louisiana, this petrified wood is also recognized as the state stone of Mississippi. The third state in the palm-wood-rich tri-state area, Texas, also designates petrified palm wood, honoring it as the state's official stone.

By now, there is little need to discuss the formation and structure of this wood. However, we shall touch upon the basics for those who have not read the preceding sections. What we call petrified wood is basically nothing more than fossilized tree stems, trunks, and branches. These pieces of wood die, are buried, and become permeated by a variety of minerals, which include silica, sulfates, sulfides, oxides, silicates, carbonates, and phosphates. This mineralized wood is basically found in two types of strata: those formed rapidly by volcanic ashfall, and those formed gradually by stream-deposited mud accumulation. The petrified wood found in Texas is found in latter type of sediment. After the wood died and was buried, dissolved silica within the groundwater attached itself to the lignin and cellulose of the wood. Over time, a layer of silicic acid formed on the exposed woody tissues. This acid would then dehydrate into a silica gel, sucking water out of the wood. Within the next several million years this silica, now known as opal, further dehydrated and reorganized into microcrystalline quartz, or chert. During this process, the aforementioned minerals and a wide variety of other materials become incorporated into the silica gel, causing it to change colors. For this reason, the petrified palm wood found in Texas comes in a variety of hues.

Of all the petrified wood found in Texas, petrified palm wood is likely the most common. This wood is basically a group of silicified woods from fossil monocots that contain prominent rod-like structures within the regular grain of the wood. Often, these structures show up as spots, tapering rods, or continuous lines. Although they have altered, these rod-like structures are remnants of sclerenchyma bundles that once comprised part of the tissue that gave the wood its strength.

The petrified palm wood found in Texas belongs to the paleobotanical form genus *Palmoxylon*. Seven distinct species of *Palmoxylon* are known from the Texas-Louisiana-Mississippi area, with the species *Palmoxylon texense* known only from Texas. Other species, such as *Palmoxylon microxylon, Palmoxylon cellulosum*, and *Palmoxylon lacunosum,* are also known from the area.

All of these trees grew in what is now Texas about 30 million years ago, during the Oligocene Epoch. During this time, the Gulf of Mexico extended much farther north, which explains why petrified wood specimens often turn up in northern areas of Louisiana, Mississippi, and Texas. While it is very difficult to deduce much about the lifestyle and appearance of these palm trees, references to modern palms are most helpful. Therefore, it is safe to assume that Oligocene palms had only a set of clustered leaves at the top of the tree. Lack of any branching nodes preserved in the wood corroborates this assessment.

Today, this petrified wood is commonly found in the Toledo Bend area, a section of land shared by both Texas and Louisiana. This area, in turn, is located on the Gulf Coastal Plain.

The Texas State Legislature officially adopted petrified palm wood as the state stone in 1969, long before a state dinosaur was recognized. The House of Representatives adopted the measure on March 19, 1969, followed by the Senate five days later. Officially, on March 26, 1969, petrified palm wood became the state stone.

Petrified palm wood is a unique and interesting type of fossil. Because of its natural toughness and wide array of colors, it is commonly collected by rockhounds. Luckily for fossil enthusiasts, an abundance of this wood remains in Louisiana, Mississippi, and Texas.

FOR FURTHER INFORMATION:

Berry, E. W., 1916. The Flora of the Catahoula Sandstone. U.S. Geological Survey Professional Paper no. 98, p.227-251.

Heinrich, P.V., Petrified Palm Wood, http://www.intersurf.com/~heinrich/Palm.html

McMackin, C. E., 1984. Petrified Wood From East to West; Some We've Liked Best. Lapidary Journal, vol. 37, no. 11, p.1582-1588.

WASHINGTON

PETRIFIED WOOD

When most people think of fossils, large dinosaurs like *Tyrannosaurus* or *Apatosaurus* frequently come to mind. However, with state legislatures, it seems as if a much different type of fossil is more popular. The most common official fossil, which has garnered six different designations (ranging from state fossil, to state gem, to state stone), is petrified wood.

While Washington recognizes the Columbian mammoth as its official state fossil, it also designates petrified wood as its state gem. This designation is more than fitting, as petrified wood is likely the most common fossil found in the entire state. Most of this wood can be traced back to the Tertiary Period.

Specifically, much of this petrified wood is known from the Miocene, an Epoch during which the state was much different. Today divided into rainy coastal and arid interior regions, Washington was swampy and mild during the Miocene, supporting a vast ecosystem of cypress, oak, elm, and ginkgo trees growing in marshy areas. These trees made their home next to large volcanoes, which would occasionally erupt, trapping countless specimens of logs inside massive ash falls and lava flows. Over time, these logs turned to stone.

The process of wood petrification should be well known to the reader of this book. Basically, what we call petrified wood is nothing more than the branches, stems, and roots of trees. This petrified wood is formed when the woody parts of a tree die and are buried. Over time, minerals from the groundwater, most often silica, permeate this wood, dehydrating it. The wood continues to dry out for millions of years, and often changes color due to the presence of foreign substances that also enter via the groundwater, until it is revealed by erosion.

Petrified wood is commonly found in two types of sediments: those formed by volcanoes and those deposited by streams. Obviously, the majority of the petrified wood found in Washington is known from the former type of sediments. Many of these sediments compose the Miocene-age Columbia Plateau Basalts, the major petrified wood-bearing unit in Washington.

Washington's most famous petrified wood is found in Ginkgo Petrified Forest State Park in eastern Washington, near the town of Vantage. This park preserves wood from Miocene-aged trees that were constantly buried in hot lava. Over time this lava would cool, and groundwater would permeate the logs, preserving them. Obviously, the most common type of petrified wood found in the park is that of ginkgos. Ginkgos, gymnosperm trees, were very common during the Mesozoic, but it was thought that they had gone extinct until living specimens were discovered in China in the last century. They are now grown under cultivation in many parts of the Far East, as well as in this country, and are called "maidenhair trees" after the resemblance of their leaves to those of the maidenhair fern.

Ginkgo Petrified Forest State Park was established by the Washington State Legislature in 1935. Today, the park is open year-round, and includes nature trails, boat docks, camping grounds, and an interpretive center, which is located in the Heritage Area of the park. This center was constructed in 1936, and exhibits over 50 varieties of cut and polished petrified wood. Most of this wood is from the park itself. Much of this wood has been exposed due to Ice Age floods, years of erosion, and sometimes excavation by humans. More specimens turn up each year. However, this park is not the only site in Washington where petrified wood is abundant.

Other good petrified wood sites are located near Saddle Mountain and along Yakima Canyon in eastern Washington. There are also well-documented sites located in western and southwestern Washington, although they have received less press than the eastern sites. Among the best sites in southwest Washington are Salmon Creek near Toledo, Silver Lake near Kelso, and Tono. Petrified wood can sometimes be found along the beaches of southwestern Washington, but is rare.

There are few published reports on the petrified wood of western Washington, despite the fact that this wood is abundant and diverse. Much of the early wood from this area was collected in the Chehalis Valley by Hugh Brown, and studied by Professor George Beck of Washington State College, one of the most famous paleobotanists of his time. Beck noted that these woods likely formed during the early Tertiary, and match fairly well with the woods found near Yellowstone National Park. Beck documented specimens of *Quercus*, the live oak, *Fagus*, the beech, and *Ulmus*, the elm, among others. Later, in the 1950s, Charles E. Isham of Vancouver, Washington conducted a study of plant fossils found near Ladd Mine.

Today, good specimens of petrified wood still turn up in southern Pierce and Thurston Counties in western Washington, as well as near the famous State Park. There has also been wood reported from the Ashford Hills, associated with the Ohanapecosh Formation.

Because of the abundance and popularity of this petrified wood, the Washington State Legislature named it the official state gem on March 12, 1975. The credit for this designation is due to many local rock groups, which lobbied for the designation. Today, because of their efforts, Washington is the sixth state to honor the ever-popular petrified wood.

FOR FURTHER INFORMATION:

Beck, G. F., 1941. Fossil Woods of the Far West. Central Washington College Geology Newsletter, vol. 1, no. 5, p. 2.

Edmondson, S. L., 1991. Petrified Wood from Washington. Rocks and Minerals, vol. 66, no. 4, p.321-325.

Pattie, R., 1983. Gems and Minerals of Washington. Washington Department of Natural Resources Pamphlet.

Strauss, E., Petrified Wood from Western Washington, http://www.mashell.com/~estrauss/pwoodfx.html

The Symbols of Washington, http://www.leg.wa.gov/legis/symbols/symbols.htm

WEST VIRGINIA

FOSSIL CORAL

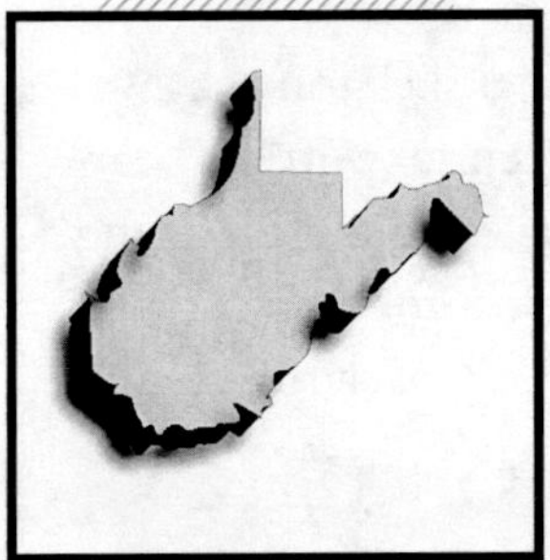

As John Denver would attest, West Virginia is one of the most beautiful states in the U.S. What Denver didn't mention in his song, though, were the fossils of West Virginia. Although the state is small in relation to other states and is characterized by granite mountains, fossils can be commonly found in West Virginia. These fossils include a plethora of marine invertebrates, some land vertebrates, and most renowned, a glut of Pennsylvanian fossil plants. These plants, among them several ferns, grew in dense coastal swamps. Today, the remnants of these swamps are the vast beds of coal that are mined throughout the state. However, despite this abundance and diversity of fossils, West Virginia has not named an official state fossil.

West Virginia has dubbed a fossil as its state gem, though. This fossil, commonly called West Virginia fossil coral, is known to science as *Lithostrotionella*. This coral lived during the Mississippian Period, about 340 million years ago, a few million years before West Virginia's vast coal fields were formed. Like all corals, *Lithostrotionella* was a member of the phylum Coelenterata, was radially symmetrical, and had an internal space used for digestion, commonly called a gastrovascular cavity. This cavity opens into a mouth, which is surrounded by tentacles. These tentacles facilitate the eating process by helping to capture small particles of food, which make up the majority of the coral individual's diet.

Corals, *Lithostrotionella* included, are immediately recognized by their stony exoskeleton, a framework consisting of calcium carbonate and secreted by each individual coral animal. This skeleton provides support for the living coral, commonly known as a polyp, and also serves as a protective enclosure into which the polyp might retreat when threatened by predators.

Corals come in two common varieties: solitary and colonial. Most corals are colonial, and many form vast reefs. *Lithostrotionella* is a member of this more common group. The *Lithostrotionella* polyp secreted a skeleton that was generally hexagonal in shape, and was characterized by a central portion bearing vertical partitions, otherwise known as septae. Often, a central post is found anchoring this exoskeleton. The hexagonal shape was perfect for forming vast colonies, and ideally enabled *Lithostrotionella* polyps to densely pack together—honeycomb-style—forming vast reefs.

During the Mississippian the lands that made up West Virginia were mostly underwater, covered by a sea that was usually warm and clear. However, the strange preservation of some corals, including those which are found fossilized in an upside-down position, seems to show that the seas could often get turbulent. However, these upside-down corals are rare, which likely means that for the most part the seas were ideal for the formation of large reefs. These reefs, in turn, supported a vast ocean ecosystem, which included brachiopods, trilobites, and fish.

However, like any animal, a coral individual would die after a while. Because of the tough calcium carbonate skeleton, the chance that a dead coral would become preserved was relatively high. After the coral died, its skeleton would become saturated with water. This water contained dissolved silica, which replaced the coral's decaying soft parts. Cavities in the coral were filled with chalcedony, a variety of quartz. Over time, as the oceans receded, hard limestone began to form around sections of coral reefs, protecting them from the elements. Today, large sections of these limestone-covered reefs are found in West Virginia.

West Virginia fossil corals come in a variety of colors, ranging from dark grey to red. The reason for this variety of colors lies in the aforementioned chalcedony. Chalcedony is a very fine-grained variety of quartz. It manifests itself with a waxy luster and appears in a wide variety of colors. In nature, it often forms rounded crusts and rinds, and is also known to form stalactites, the familiar dripping structures seen in caves. As a rule, chalcedony has a compact fibrous structure and fine splintery fracture, properties that make it ideal for preserving petrified wood and fossil corals.

While chalcedony petrified wood is common in the American southwest, the best chalcedony fossil corals are likely those found in West Virginia. These specimens are so common that in many places farmers have been known to throw large chunks of coral down sinkholes to clear them from their fields. Some pieces of this coral, known simply as "coral heads," can weigh as much as 15 pounds. Although coral heads are much rarer than smaller specimens, all types of coral can be commonly found in rock quarries, road cuts, and plowed fields. This coral is abundantly sold in West Virginia shops and over the Internet, with the denser specimens garnering the higher prices.

Lithostrotionella is almost exclusively found in West Virginia's Hillsdale Limestone, a Mississippian unit that occurs in Greenbrier and Pocahontas counties in the southeastern portion of the state. The specimens are so common that they are commonly seen in local jewelry stores, and are sold as gems. Because of this, the West Virginia State Legislature passed House Concurrent Resolution Number 39 on March 10, 1990, officially designating the West Virginia Fossil Coral (*Lithostrotionella* sp.) as the state gem.

Although West Virginia is famous for its well-preserved Pennsylvanian plants, the state has not designated an official state fossil. Instead, West Virginia has recognized a common Mississippian coral as its state gem. Luckily, for collectors, this "gem" can be commonly found both in the field and in souvenir shops.

FOR FURTHER INFORMATION:

Pellant, C., 1996. An Illustrated Guide to Fossils. The Nature Company, London.

West Virginia Geological and Economic Survey, 1959. Common Rocks and Minerals of West Virginia. Publication ED-7.

PROPOSED STATE FOSSILS

PROPOSED STATE FOSSILS

While most states have designated state fossils, several have chosen not to (or just haven't gotten around to it). However, many of these states have proposed state fossils at one point or another. For various reasons these fossils either failed in the State Legislature or are still in committee. Although it is impossible to list every fossil recommended for state honors, here are descriptions of a select few proposed recently. Perhaps, if you live in one of these states, you or a group you are associated with might take up the cause?

ARKANSAS—*Arkansaurus fridayi*

A few years ago, paleontologist John McFarland conducted a survey of various Arkansas geologists, asking them what fossil they thought best represented the state. Many petroleum geologists recommended tiny microfossils, such as foraminifera and conodonts, because of their importance to the oil industry. Others suggested a plant to represent Arkansas' coal deposits, while a large group suggested a dinosaur: *Arkansaurus fridayi.* Unfortunately, this dinosaur is poorly known, as only select bones from its right foot have been found.

Arkansaurus, which literally means "Arkansas lizard," was a bird-like bipedal theropod that lived during the Late Cretaceous. It is only known from fossils found in Arkansas, but is related to several coelurosaurs, and is possibly an ornithomimid. The only specimen of *Arkansaurus*, the only known dinosaur from Arkansas, was found near Lockesburg by J.B. Friday during the construction of a road in 1972. As the story goes, Friday was looking for a lost cow on his farm when he stumbled across the construction site, where he proceeded to find the bones. Friday showed his bones to a University of Arkansas professor, James Quinn, who identified them as belonging to a dinosaur. He suggested that casts be made of the specimen, and later three were molded. They are currently on display at the University of Arkansas in Fayetteville, the Arkansas Geological Commission, and the Science and History Museum at MacArthur Park in Little Rock.

Quinn mentioned the dinosaur in a 1973 paper, but the specimen has never been properly described. Technically, then, it is a nomen nudum. Unfortunately, there is also little hope for the specimen ever being described. While it is well-documented that the find consisted of three metatarsals, three phalanges, and two claws, all records of the original fossil's whereabouts have disappeared. According to most records, the bones were supposed to be housed at the University of Arkansas, but they are nowhere to be found. This has led some to believe that *Arkansaurus* was nothing more than a hoax, craftily created by molding plaster into the shape of a generalized theropod foot. However, these claims are unsubstantiated. Little hope of reconstructing this dinosaur remains, without finding more specimens, leaving scientists puzzled for the present.

FLORIDA—*Eupatagus antillarum*

As mentioned earlier in the text, Florida is home to a vast and diverse fossil record. However, the state has yet to designate an official fossil. Florida does have an unofficial state fossil, though—the echinoderm *Eupatagus antillarum.*

Eupatagus is an extinct type of sand dollar that reached sizes of four to five inches (12 centimeters) in diameter. It is an echinoderm, related to starfish and crinoids, and possessed a five-fold symmetry and a spiny-type skin. *Eupatagus* dates from the Eocene Epoch, which lasted from about 35-60 million years ago.

According to most scientists, *Eupatagus* probably lived in shallow, horizontal burrows on the bottom of the sea. It fed on tiny animals, which it received from the water via a narrow, vertical tunnel. *Eupatagus* made its home in the vast shallow seas that covered Florida during the Eocene. Today, *Eupatagus* is commonly found in the Ocala Limestone, which is subdivided into three distinct sub-formations: the Williston, Inglis, and Crystal River units.

The Florida Geological Survey nominated *Eupatagus* as the state fossil, largely because it is so familiar and very popular with collectors. The Florida Museum of Natural History in Gainesville followed their lead, creating an exhibit promoting *Eupatagus* as the "state fossil of Florida." However, the State Legislature has never followed suit, keeping this fossil unofficial for the time being.

INDIANA—*Cyathocrinites*

The crinoid *Cyathocrinites multibrachiatus* was once proposed as the official Indiana state fossil. However, the bill was sunk in the Legislature. Crinoids are common in the limestones of Indiana. Described earlier in this book, crinoids are echinoderms that look much like plants. They are filter feeders that are sessile, with skeletons composed of hard, tough ossicles. These ossicles are commonly found fossilized in limestone.

IOWA—crinoid

The crinoid has also been suggested as the official state fossil of Iowa. In 1999 House Joint Resolution 8 was proposed into the Iowa State House of Representatives. To date, this measure, which would have designated the crinoid the official state fossil, has failed. However, it is still active, and may possibly be passed soon. Also included in this bill is a passage stating that the citizens of Iowa are "encouraged to tour the current exhibit at the State Historical Building entitled 'Flowers of the Iowa Seas' which includes specimens of fossil crinoids." Many of the crinoids in this exhibit were found in the LeGrand and Burlington area of Iowa. Other fossil crinoids are found in Marshall County. Perhaps the state's most well-known crinoid-bearing unit is the Keokuk Formation, a Mississippian rock unit that preserves a variety of crinoids.

KANSAS—*Xiphactinus*

Since the spring of 2000, Alan Detrich, an often-eccentric fossil hound who once tried to peddle a *Tyrannosaurus rex* skeleton for $20 million, began to promote the large, carnivorous fish *Xiphactinus* as Kansas' official state fossil. This large fish, which lived during the Late Cretaceous, grew to sizes of 18-20 feet. It is commonly found in Kansas, a state that can boast one of the most extensive Cretaceous marine deposits known in the world.

In order to help his cause, Detrich enlisted several Kansas grade schools to lobby for *Xiphactinus*' designation. He even urged voters in the November 7 election to write in the name of the fish on ballots where local politicians were running uncontested. Election officials complained of this, and Detrich decided to withdraw his plea. However, he only agreed to stop the write-ins in exchange for pledges of support from several local candidates.

Detrich plans to have a state representative endorse the cause, and introduce the *Xiphactinus* bill to the State Legislature. No bill has yet been proposed, though.

MICHIGAN—mastodon

Michigan has designated a fossil as its official state stone, the coral *Hexagonaria*, otherwise known as the Petoskey Stone. However, Dave Thomas, a geology instructor at Washtenaw Community College, is now working to convince state legislators to name the mastodon the official state fossil. Surprisingly, this is causing much controversy.

Despite the fact that Thomas found an ally to support the bill, Senator Thaddeus McCotter from Livonia, the legislation is drawing much heat. While the bill was introduced to the State Senate in April, 2001, several legislators, and even the Governor, do not support it. Governor John Engler believes a limit should be set on the seemingly-endless official state symbols, and it looks as if he may veto the bill if it does indeed pass. There is no word on its future yet.

Thomas, who worked on a mastodon dig in Saline, Michigan, believes that the designation of a state fossil can be used by science teachers around the state as a teaching tool. Michigan boasts the world's best-known set of mastodon footprints, and fossil remains of this giant elephant are commonly found in 10,000-20,000-year-old sediments.

After his plea to use the mastodon as a teaching tool, Thomas received a $300 grant from the Michigan Earth Science Teachers Association to fund his cause. He has even decided to run a statewide contest for school children that will award prizes to students who write the best letters lobbying their state representatives. Children historically have much clout when it comes to issues involving state fossils, so Bill 397 may indeed be passed, giving Michigan its own state fossil.

MINNESOTA—*Castoroides* and *Rynchotrema*

Minnesota has not designated any fossils as official symbols, but it has twice tried to designate a state fossil. Both failed designations came in 1988, with the giant beaver *Castoroides ohioensis* and the articulate brachiopod *Rhynchotrema* failing. The giant beaver, proposed by bill SF1701, was sponsored by State Senator Larry Pogemiller. This large mammal was suggested by third grade students at North Star Primary School, but ultimately failed in its bid to become an official fossil.

PENNSYLVANIA—*Atreipus milfordensis*

It looks as if Pennsylvania will soon be able to claim an official state dinosaur. In addition to its state fossil, the trilobite *Phacops rana,* Pennsylvania will likely designate the dog-sized herbivorous dinosaur *Atreipus milfordensis* as an official symbol. This small dinosaur, well-known from tracks in the Pennsylvania area, was about three to five feet long, walked on all fours, and lived during the Late Triassic Period. Many paleontologists believe that it travelled in flocks across the eastern United States.

A bill to nominate *Atreipus* was introduced in Pennsylvania's State Legislature in May of 2001. Peter Kranz, a free-lance Maryland fossil hunter who previously lobbied for the designation of Maryland's state dinosaur, is the driving force behind this bill. However, like the mastodon bill of Michigan, the legislation making *Atreipus* the official dinosaur is also controversial. State Representative Jim Wansacz, from Lackawanna, lamented that he grew up in a blue-collar district, and didn't want a "wimpy dinosaur." Despite his likely regrets, the *Atreipus* bill is expected to pass.

RESOURCES

The following resources, not mentioned in the text, are helpful for those hoping to find out more about the state fossils and other aspects of paleontology:

Fossil News: Journal of Avocational Paleontology
http://www.fossilnews.com
Monthly magazine for those who pursue paleontology as a hobby.

Prehistoric Times
http://www.prehistorictimes.com
Bi-monthly magazine with a focus on paleoart and interviews.

Dino Land
http://www.geocities.com/stegob
My own website, focusing on all aspects of paleontology, including interviews, photos, and original articles.

Dino Land State Fossil Page
http://www.geocities.com/stegob/statefossilpage.html
My own web page, which focuses on the state fossils.

Dino Land Dinosaur Gazette E-Mail Newsletter
Subscribe at: brusatte@theramp.net
Monthly e-mail newsletter featuring news articles, book reviews, and original fossil articles.

Dinosaur Interplanetary Gazette State Fossil Page
http://www.dinosaur.org/statefossil.htm

Credits

PHOTOGRAPHS (fossil / photographer / location)

p. 8: *Basilosaurus*/Stephen Brusatte/Dinofest 2000, Chicago
p. 12: *Mammuthus* sp. tooth/Lynne M. Clos/Denver Museum of Nature & Science
p. 14: Woolly mammoth reconstruction/Lynne M. Clos/George C. Page Museum of La Brea Discoveries
p. 17: *Araucarioxylon arizonicum*/Lynne M. Clos/Denver Museum of Nature & Science
p. 19: *Araucarioxylon arizonicum*/Lynne M. Clos/private collection
p. 23: *Smilodon californicus*/Stephen Brusatte/New Mexico Museum of Natural History
p. 24: *Smilodon californicus*/Lynne M. Clos/George C. Page Museum of La Brea Discoveries
p. 28: *Stegosaurus*/Lynne M. Clos/Natural History Museum of Los Angeles County
p. 30: *Stegosaurus stenops* (with juvenile and *Allosaurus*)/Lynne M. Clos/ Denver Museum of Nature & Science
p. 33: *Eubrontes* track/Stephen Brusatte/Dinosaur State Park
p. 35: *Eubrontes* track/Stephen Brusatte/Dinosaur State Park
p. 38: *Belemnitella americana*/Stephen Brusatte/Field Museum of Natural History
p. 43: *C. megalodon* teeth/Marc Behrendt/private collection
p. 44: *C. megalodon* tooth/Marc Behrendt/private collection
p. 47: *Equus simplicidens*/Stephen Brusatte/National Museum of Natural History (Smithsonian)
p. 48-49: *Equus simplicidens*/Lynne M. Clos/Natural History Museum of Los Angeles County
p. 53: *Tullimonstrum gregarium*/Stephen Brusatte/Field Museum of Natural History

p. 54: Tully Monster model/Stephen Brusatte/Burpee Museum of Natural History

p. 57: brachiopods/Stephen Brusatte/National Museum of Natural History (Smithsonian)

p. 58: brachiopods/Lynne M. Clos/Colorado Fossil Expo 2001

p. 62: petrified "palm" wood/Stephen Brusatte/Dallas Museum of Natural History

p. 78: *Grallator* track/Stephen Brusatte/Gary Gaulin footprint site

p. 79: Author and Gary Gaulin/Roxanne Brusatte/Gary Gaulin footprint site

p. 80: *Eubrontes* track (natural cast)/Lynne M. Clos/Museum of Western Colorado's Dinosaur Journey

p. 83: *Zygorhiza* reconstruction/Lynne M. Clos/Las Vegas Natural History Museum

p. 85: *Basilosaurus* skull/Stephen Brusatte/National Museum of Natural History (Smithsonian)

p. 93: *Maiasaura*/Stephen Brusatte/Dinofest 2000, Chicago

p. 98: *Mammuthus columbi*/Lynne M. Clos/George C. Page Museum of La Brea Discoveries

p. 99: *Mammuthus columbi*/Allen A. Debus/Geology Museum, Salt Lake City

p. 102: *Shonisaurus* bones/Allen A. Debus/Berlin-Ichthyosaur State Park

p. 103: *Shonisaurus* reconstruction/Lynne M. Clos/Las Vegas Natural History Museum

p. 107: *Coelophysis bauri*/Lynne M. Clos/Natural History Museum of Los Angeles County

p. 108: *Coelophysis bauri*/Lynne M. Clos/Denver Museum of Nature & Science

p. 112: *Eurypterus remipes*/Lynne M. Clos/Denver Museum of Nature & Science

p. 117: modern *Teredo* wood/Stephen Brusatte/Virginia Living Museum

p. 121: *Isotelus* model/Stephen Brusatte/National Museum of Natural History (Smithsonian)

p. 122: *Isotelus latus*/Marc Behrendt/private collection

p. 126: *Saurophaganax maximus*/Stephen Brusatte/Dinofest 2000, Chicago

p. 131(both): *Phacops rana*/Marc Behrendt/private collection

p. 132: *Phacops rana*/Lynne M. Clos/Denver Museum of Nature & Science
p. 133: *Phacops rana*/Marc Behrendt/private collection
p. 136: *Triceratops* skull/Lynne M. Clos/Natural History Museum of Los Angeles County
p. 141: *Pterotrigonia thoracica*/Stephen Brusatte/National Museum of Natural History (Smithsonian)
p. 142: *Pterotrigonia thoracica*/Stephen Brusatte/National Museum of Natural History (Smithsonian)
p. 144: *Allosaurus fragilis*/Lynne M. Clos/Museum of Western Colorado's Dinosaur Journey
p. 150: *Delphinapterus leucas*/Stephen Brusatte/Redpath Museum
p. 154: *Chesapecten jeffersonius*/LeGrand Smith/private collection
p. 156: *Chesapecten jeffersonius*/Stephen Brusatte/National Museum of Natural History (Smithsonian)
p. 159: *Mammuthus columbi* skull/Lynne M. Clos/Denver Museum of Nature & Science
p. 163: *Calymene (Gravicalymene) celebra*/Lynne M. Clos/Denver Museum of Nature & Science
p. 164: *Calymene celebra*/Teddie Mobley/private collection
p. 165: *Calymene celebra*/Teddie Mobley/private collection
p. 168-69: *Knightia* mass burial/Lynne M. Clos/Denver Museum of Nature & Science
p. 171: *Knightia* set into oak box/Lynne M. Clos/private collection
p. 177: *Astrodon johnstoni* tooth/Stephen Brusatte/Dinofest 2000, Chicago
p. 181: *Hadrosaurus* sculpture by Allen A. Debus/Allen A. Debus/private collection
p. 186: *Pleurocoelus* vertebra/Stephen Brusatte/Ft. Worth Museum of Science & History
p. 187: *Brontopodus* track/Stephen Brusatte/Dallas Museum of Natural History
p. 198: "agate" on coral head/Lynne M. Clos/private collection
p. 202: Petoskey stone/Teddie Mobley/private collection
p. 206: petrified "palm" wood/Stephen Brusatte/Dallas Museum of Natural History

p. 209: petrified "palm" wood/Stephen Brusatte/Dallas Museum of Natural History
p. 212: petrified wood/Lynne M. Clos/private collection
p. 217: Appalachian fossil coral (gen. sp. indet.)/Lynne M. Clos/private collection
p. 232: Author with *Triceratops* skull/Roxanne Brusatte/National Museum of Natural History (Smithsonian)

DRAWINGS

pp. 13, 27, 44-45, 104, 146, 192, 235: Shannon S. Yeager
pp. 68, 72, 88, 97, 183: Lynne M. Clos
pp. 164, 188, and state maps: Corel Gallery

About the Author

Stephen Brusatte is an undergraduate geophysics student at the University of Chicago. Steve has authored over 100 paleontology articles and papers, and has been published in *Fossil News, Dino Press, Dinosaur World, Prehistoric Times*, and several other publications. He maintains his own paleontology website, Dino Land <http://www.geocities.com/stegob>, and during the summer works as a sports reporter for the Ottawa Daily Times. His major scientific work thus far has been a description of Upper Pennsylvanian brachiopods from the Illinois Basin, which he has used to infer the paleoecology of Central Illinois 280-300 million years ago. This work was awarded a prize in the National Junior Science and Humanities Symposium in 2002. Born and raised in Central Illinois, Steve splits his time between Chicago and Ottawa, Illinois.

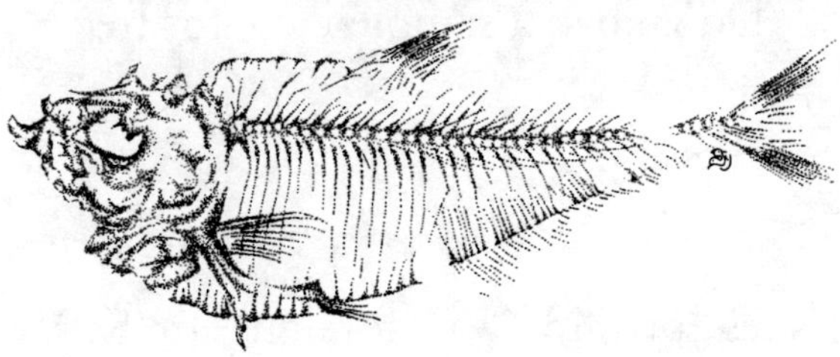

Yes! Sign me up for Fossil News:

☐ One year printed $43

☐ Two years printed $80 (includes CD)

☐ One year printed (outside U.S.) $55 U.S.

☐ One year online $10

☐ Two years online $20 (includes CD)

Email (if online subscription): ______________________________

Name ______________________________

Address* ______________________________

____

City, State/Province ______________________________

Zip/Postal Code, Country ______________________________

☐ Check payable to Fossil News enclosed.

☐ Charge my (circle) Visa Mastercard Amex Discover

Card number ______________________________

Expiration date ______________

Mail to: Fossil News, 1185 Claremont Dr., Boulder, Colorado 80305 USA

*Please include snail mail address even if online subscription so we can contact you in case of a problem with email.